P.R. Jones

Identification
of Organic Compounds
with the Aid
of Gas Chromatography

Identification of Organic Compounds with the Aid of Gas Chromatography

RAYMOND C. CRIPPEN, Ph.D.

Industrial Consultant
Director of Laboratories: Mathematics and Science Department
Northern Kentucky State College

McGRAW-HILL BOOK COMPANY

New York   St. Louis   San Francisco   Düsseldorf   Johannesburg
Kuala Lumpur   London   Mexico   Montreal   New Delhi
Panama   Rio de Janeiro   Singapore   Sydney   Toronto

Library of Congress Cataloging in Publication Data

Crippen, Raymond C.
  Identification of organic compounds with the aid of
gas chromatography.

  Includes bibliographical references.
  1. Gas chromatography.   2. Chemistry, Analytic—
Qualitative.   3. Chemistry, Organic.   I. Title.
QD271.C768   547'.34'926   73–4966
ISBN 0–07–013725–0

1234567890KPKP76543

*The editors for this book were William G. Salo, Jeremy Robinson, and
Lila M. Gardner, the designer was Naomi Auerbach, and its production
was supervised by Teresa F. Leaden.   It was set in Imprint 101.*

It was printed and bound by The Kingsport Press.

This book is dedicated to my wife, Helen,
whose encouragement and assistance made this book possible
and to the memory of Joy Derickson,
a brave and courageous girl who really wanted to learn Chemistry.

Contents

Preface

Thanks to the rapid development of gas-chromatographic instruments and techniques, this innovation in the field of analytical chemistry can both measure the quantity of a component and give an indication of its identity. The gas chromatograph has recently been exploited for the identification of organic components. The retention time, relative retention time, and other values used to measure the time of elution of a given component on a specific column under a standard set of conditions are helpful in establishing the identity of the component. However, since many components may have similar elution times, the behavior must be studied on more than one chromatographic column.

After a component has been eluted from columns with nonpolar, medium-polar, and highly polar properties, only conversion to a variety of derivatives and remeasuring on various columns will really help verify the component's identity. If absolute identity is required, the component can be collected on a preparative gas chromatograph and examined by other instrumental means.

This book carries the investigator through the various standard steps necessary to establish the identity of a substance with the continuous assistance of the gas chromatograph. It shows how relatively non-volatile derivatives can be made volatile by further conversion to trimethylsilyl groups, $-Si(CH_3)_3$, or other more volatile derivatives, where unknown compounds have active hydrogens.

Exhaustive literature studies have not been attempted on every facet of this subject. When an article partially or completely illustrates the point, it alone is cited. Where no literature was available, original research was performed to develop the data.

Since retention data are more effectively presented as a series of curves than as tables, all retention data are either presented as the direct chromatogram or are plotted as a homologous series against some linear physical property, such as carbon number, boiling point, or refractive index. Solubility data have been shown to be logarithmic properties.

The author is indebted to personnel of Richardson-Merrell, Inc., Cincinnati, Ohio, for assistance in writing this manuscript and to various local firms for use of their equipment and supplies to develop the data required. The author also appreciates the assistance of R. S. Benner, Aquaphase Laboratories, Adrian, Mich., who made valued suggestions. The encouragement of friends, business associates, and relatives and especially the patience and help of my wife, Helen, during the many hours of preparing this book are greatly appreciated.

Raymond C. Crippen

Cincinnati, Ohio

Identification
of Organic Compounds
with the Aid
of Gas Chromatography

CHAPTER ONE

Introduction

Every substance found in nature or synthesized by man has impurities to a greater or lesser extent. These unknown substances must be identified in order to reduce or remove the impurities, and increase the yields.

The gas chromatograph is an efficient tool for separating mixtures of compounds into their individual components (see Fig. 1-1). Since the separated compounds "do not come out labeled," it becomes the task of the analytical chemist to establish the identity of these separated substances.

GENERAL IDENTIFICATION OF ORGANIC COMPOUNDS

This book explains how the chemist can separate various types of compounds using a variety of analytical gas-chromatographic columns. Preliminary separation by solubilities in suitable solvents and reagents and partial fractional crystallization or fractionation make the analysis

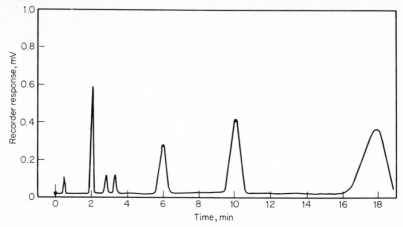

Fig. 1-1 Gas chromatographic separation of components of a lacquer solvent ("they do not come out labeled ").[6]

easier, and determination of funtional groups also helps the analyst. Special problem separations require special techniques.

In identifying the separated components, many other procedures, in addition to the regular techniques, will be used, e.g., retention times versus those of known compounds on the same column, additive and subtractive techniques, and the use of derivatives.   Most literature on known compounds will be verified by these techniques.   When compounds unknown in the literature are separated, they must be collected and examined by other instrumental methods, e.g., infrared, ultraviolet, nuclear magnetic resonance, polarography, Raman spectroscopy, chemical microscopy, and others.   Cheronis and Entrikin[3] also reported a classification by indicators.

As much as possible should be discovered about the physical and chemical composition or properties of each sample, e.g., solubility, melting or boiling point, density, elemental composition, and functional groups.   If the substance is impure, these physical and chemical properties should be determined on a purified portion of the sample. Chapter 4 will show how physical properties can be predicted from retention data.[9]

Color, odor, and other physical properties often give a clue to the composition of a sample.   Elemental analysis should be performed routinely on all samples.   When nitrogen, halogen, sulfur, phosphorus, or other elements are found besides carbon, hydrogen, or oxygen, it is a foregone conclusion that a functional group containing one of these elements is present.

It is helpful to know what use was or will be made of the sample.

Further clues to its probable composition come now knowing how other similar preparations were formulated or synthesized.

Samples may be reported on forms of the type shown by Shriner, Fuson, and Curtin[11] or various commercial forms. As described later in this chapter, a notebook should be kept recording all tests and results. Even the most minor observations on a physical or chemical property will give the chemist some indication of the probable composition of the unknown.[8]

This book will acquaint the chemist with the versatile technique of gas chromatography generally and gas-liquid chromatography specifically,[7] but it should not be expected that these procedures will supplant all others. The chemist should learn to use these techniques in conjunction with other standard procedures for chemical identifications and problem solving. Each instrument and technique has its applications and limitations; where one is not applicable, another should apply.[10]

Instead of including numerous tables of melting points, retention data, and other physical and chemical properties the data have been presented in graphic form for easy retrieval. Although there is a loss in accuracy, it can be retrieved by replotting on a wider scale.

IDENTIFICATION OF NEW COMPOUNDS

It should be relatively easy to verify the composition of known or probable mixtures by means of gas-liquid chromatography alone, but in completely unknown samples, one must utilize infrared, nuclear magnetic resonance, ultraviolet absorption, or other instrumental and

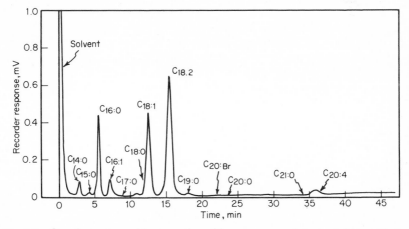

Fig. 1-2 Gas-chromatographic separation of human serum cholesterol fatty acids (as methyl esters) on ethylene glycol succinate column at 185°C.[5]

chemical techniques to verify the identity of a given component.   An unknown may behave on several chromatographic columns as a suspected known, but until the analyst has actually collected the unknown and examined it by other instrumental means, its exact composition has not been determined.   Since two different compounds may have the same polar or nonpolar properties, other means of analysis are required for the exact identification (see Fig. 1-2).   A trace component in a fatty acid chromatogram may be a branched or unsaturated fatty acid.

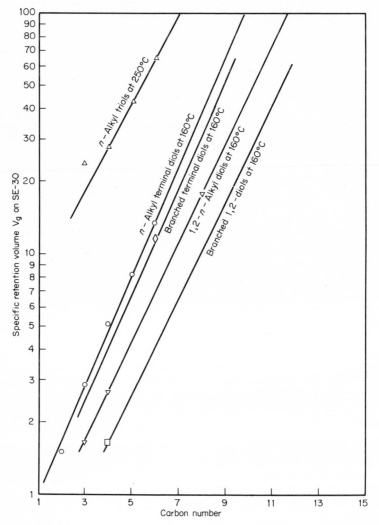

Fig. 1-3   Separation of alkyl diols and triols on silicone gum rubber (SE-30) at 160 and 250°C.[9]

When the physical and chemical properties do not match the information found in the literature, it is necessary to synthesize the compound and match its properties to the separated substance (see Fig. 1-3). In this example, the infrared, ultraviolet, near-infrared, and nuclear magnetic spectrometric patterns indicated that no terminal methyl groups were present on a five-membered carbon chain with two hydroxyl groups. Gas-chromatographic retention data predicted a probable structure of 1,5-dihydroxypentane. However, verification could not be obtained until actual synthesis of the compound was carried out.

GAS-CHROMATOGRAPHIC EXAMINATION TECHNIQUES

By itself, gas chromatography can separate individual components from complex mixtures and give a measure of the quantity of each component present. When used with other physical and chemical measurements, it assists materially in identifying the isolated component.

Gas chromatography gives information relative to the solubility, molecular weight, melting or boiling point, density, refractive index, and presence or absence of functional groups on relatively pure compounds (see Chaps. 2 to 4).

Samples injected into the gas chromatograph fall into three categories: (1) volatile and stable during analysis; (2) volatile and nonstable during analysis; or (3) nonvolatile (stable or nonstable).

Most materials fall into category 1 without further analytical preparation. The materials in category 2 must be converted into stable and volatile compounds by suitable treatment; e.g., ascorbic acid, which decomposes easily, is derivatized to a trimethylsilyl compound that makes it stable though volatile (see Fig. 1-4). Many compounds that are nonvolatile and do not decompose are converted into volatile derivatives so that suitable chromatograms can be obtained.

The most useful everyday application uses the gas chromatograph for qualitative analysis. The sample is injected into the instrument under known conditions, and the retention time is precisely measured. The retention time of a pure standard is measured, and its ratio to the unknown sample gives a value defined as the *relative retention time*. Comparison of this value with known retention or relative retention values and a study of the behavior on various columns (from polar to nonpolar) will give an indication of the presence or absence of functional groups. These groups can be verified by suitable reactions and further gas-chromatographic examinations (see Chap. 5). Tables of retention data will be found in the literature.[1,9]

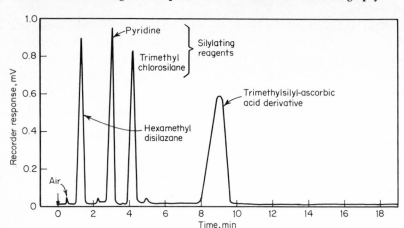

Fig. 1-4 Separation of heat-decomposable ascorbic acid by conversion to a stable trimethylsilyl derivative on 5 ft. × ⅛ in. SE-30 (10%) column at 100°C.

Solubility studies are extended by the gas-chromatographic examination of the solvent reactants.   Thus, a sample that shows relatively little solubility in the test solvent by the usual solubility tests should indicate the presence or absence of a functional group, e.g., relatively insoluble hydroxy-stearic acid may show greater solubility due to the hydroxyl than ordinary stearic acid when examined more carefully by gas chromatography.

Examination of the physical properties of the compound and comparison with homologs will make predicting the probable identity of the substance easier.   The technique is also useful in predicting the physical properties when insufficient sample is available for measurement under routine analyses or when no reliable data are available (see Chap. 4).

All the information from solubility and physical measurements should be assembled along with the results of the gas-chromatographic examination.   Using these data, the analyst can convert the sample into various derivatives and reexamine it for physical properties and gas-chromatographic behavior.   In most instances, this should serve to verify the composition of the unknown.   Finding identical retention times on three different columns (unknown versus a known) is considered good verification technique.[2]

It is essential that the elemental analysis of the substance be made early in the investigation. For example, a neutral nitrogen compound may have properties similar to a neutral hydrocarbon and may be mistaken for a hydrocarbon until it is examined by other instruments or a chemical reaction reveals the difference.

Useful combinations of instruments result when the gas chromatograph is coupled with optical instruments, such as infrared and ultraviolet units, or electromagnetic instruments, such as nuclear magnetic resonance and mass spectrometers. Suitable samples should be collected on the gas chromatograph for subsequent examination by near-infrared, ultraviolet, visible, infrared, nuclear-magnetic, or mass-spectrometric techniques. Samples can also be examined by fluorometric instruments for further verification of their ultraviolet properties.

HOW TO USE THIS TEXT

This book has been designed as a text or supplement to standard procedures of organic analysis and identification. Gas-chromatographic techniques greatly speed up the identification of unknown substances.

The chemist should refer to this volume during any qualitative organic analysis. The homologous series, as explained in these pages, should be used wherever possible; however, there are situations where the homologous series will not exactly apply, e.g., steroids. A prediction will be possible by applying these data to the nearest homologous series:

1. Using the combined techniques of gas chromatography and qualitative organic analysis, the investigator should make a preliminary test to determine whether the sample volatilizes in the instrument and generally examine it for color, odor, and other obvious physical or chemical properties. An ignition test will also give additional data, especially when the volatiles are directed into the gas chromatograph.

2. Next the sample should be examined for elemental composition. This comprises the analytical techniques available for carbon, hydrogen, and nitrogen. Alternate qualitative methods may be utilized, and sulfur, halogen, and phosphorus determination should also be made. Metallic elements will be detected in the ash residue. Fusions with elemental sodium, magnesium and sodium carbonate, zinc and sodium carbonate, or calcium oxide will give qualitative tests for many additional elements.

3. The sample should then be examined for solubility in various solvents, and each fraction is further examined on the gas chromatograph. Low solubilities may indicate possible functionality which may differ from the predicted values only slightly, e.g., 0.1 versus predicted 0.01 percent. The solubility and retention data permit the investigator to make further predictions of the identity of the substance.

4. The compound can now be examined for physical properties, which are then plotted versus the retention data in any homologous series. Many new clues to the probable identity of the substance will be uncovered.

5. The substance is now checked for possible functional groups. Each reaction should also be examined on the gas chromatograph, which sometimes uncovers a slight functionality, such as long-chain hydrocarbon versus single hydroxyl, not observable by the usual standard examinations. A plot of the reaction product versus higher and lower homologs will substantiate the possible identity.

6. At this point the analyst should prepare a series of derivatives to corroborate his tentative conclusions. If the observations are correct, the predicted retention data should correlate well with the actual observed data, including the observed melting points, boiling points, or other physical measurements.

7. On certain nonhomologous compounds (new ones not reported in the literature) the chemist may have to collect samples by trapping and examine each by other instrumental means.

8. The gas-chromatographic method will be useful for predicting many thermodynamic properties, in addition to techniques for the identification of organic compounds.

Investigators are encouraged to look at their data critically, using the techniques outlined. Some of the existing published data may thus be subject to question. For example, many of the discrepancies in the data for siloxanes were not apparent until they were examined critically by the prescribed technique. Predictions beyond the data obtained should be made by extrapolating existing curves well above the measured ranges.

NOTEBOOKS AND REPORTS

Each laboratory and each investigator has his own procedures for notebooks and report forms, but entries in all notebooks and reports must cover several primary points: (1) all observations and conclusions should be recorded as soon as they are made. Thus, the notebook will give the data chronologically as they are evolved. (2) The report differs from the notebook in that it summarizes the work recorded with emphasis on the conclusions rather than chronology.

The following information should be entered in the notebook for each known or unknown sample as it is examined:

1. Record the date the sample was received and give an indication of the nature of the sample, e.g., a single-class solvent, a mixture as "mixed solvent," or "general unknown."

2. Record the results of the gross examination, elemental analysis, solubility observations, physical measurements, and gas-chromatographic data. List the most probable compounds or class of compounds.

3. List the functional-group tests made and observations for each.

List the behavior of each reaction product in the gas-chromatographic column.

4. Summarize the probable composition of the substance based on the physical and chemical data. Compare with the data found in tables of physical properties.

5. List the types of reactions to be used for producing derivatives of the probable compound and predict how it should behave on various chromatographic columns.

6. Summarize the methods to be used to prepare the derivatives, giving typical reaction equations. Outline the methods of purification for each derivative with reasons. Record observations on melting points along with the measurements of chromatographic data.

7. Compare the data on the physical constants of the original sample and derivatives with those taken from tables or from known compounds. If necessary, compare actual and predicted data by plotting chromatographic results of homologs versus their physical constants (see Chap. 4).

8. State the final conclusions about the identity based on these observations. Verify by comparing other instrumental observations with pure known compounds. All data should agree.

The analyst can utilize any standard forms or design his own. Some investigators have adapted punched cards for their observations, such as the gas-chromatographic data cards, infrared cards, and nuclear-magnetic-resonance data cards. Gas-chromatographic literature can also be recorded on punched cards. Marking samples is important: as much information as possible should be given on each sample to simplify the identification.

The reader is referred to well-known texts for more detailed theoretical background on the subject of gas chromatography.[7,10]

REFERENCES

1. ASTM: *Compilation of Gas Chromatographic Data*, STP-343, Philadelphia, 1963.
2. ASTM: *Fall Meeting, Cincinnati, Ohio, September, 1968*, Committee E-19 on Gas Chromatographic Practice.
3. Cheronis, N. D., and J. B. Entrikin: *Semi-micro Qualitative Organic Analysis*, 2d ed., chap. 8, pp. 207–216, Interscience Publishers, New York, 1957.
4. Crippen, R. C., and C. E. Smith: *J. Gas. Chromatogr.*, **3**: 37 (1965).
5. Horning, E. C., A, Karmen, and C. C. Sweeley: *Gas Chromatography of Lipids*, p. 210, The Macmillan Company, New York, 1964.
6. Hundert, M. B.: *J. Paint Technol.*, **40**: 330–348 (1968).
7. Knox, J. H.: *Gas Chromatography*, pp. 11–38, Methuen & Co., Ltd., London, 1962.
8. McGookin, A.: *Qualitative Organic Analysis and Scientific Method*, chap. 5, p. 75, Reinhold Publishing Corporation, New York, 1955.

9. McReynolds, W. O.: *Gas Chromatographic Retention Data*, p. 44, Preston Technical Abstracts Company, Evanston, Ill., 1966.

10. Purnell, H.: *Gas Chromatography*, pp. 233–405, John Wiley & Sons, Inc., New York, 1962.

11. Shriner, R. C., R. C. Fuson, and D. Y. Curtin: *The Systematic Identification of Organic Compounds*, p. 11, John Wiley & Sons, Inc., New York, 1959.

CHAPTER TWO

Preliminary Examination and Practical Gas-chromatographic Techniques

The first step in any preliminary examination and identification of an organic compound is to determine whether it is a mixture or a pure material. If the sample is a solid, a melting point will give an indication of the degree of purity; if it is a liquid, its boiling point must be measured to ascertain the degree of purity. A faster method is to examine the material in a gas chromatograph. As described in Chap. 1, the material will fall into one of three categories: (1) it is volatile in the gas chromatograph; (2) it decomposes in the instrument; or (3) it is nonvolatile. Solids can be dissolved in suitable solvents and examined if care is taken to ensure that the solvents do not obscure the substance or contain impurities which would complicate the identification.

An unknown solid can also be examined under the microscope or magnifying glass for signs of nonhomogeneity. Crystals or particles can be separated by appearance under a lens and examined individually. If the material appears homogeneous but has a wide melting range and several peaks on the chromatogram, it can be recrystallized one or more times and examined again.

An unknown liquid may be distilled or separated by preparative gas-chromatographic procedures. If the sample is distilled, the first and last fractions can be examined by the instrument for impurities, or the refractive index can be measured on each and compared to the main body of the sample. If all fractions are nearly the same, the material may be considered relatively pure. A continuous or even stepwise rise in the distillation indicates that the material is a mixture. Technical or commercial materials give initial rises at the beginning or end, with the majority of the product distilling over at a single temperature.

CRYSTALLIZATION

The separation of solid crystals from a solution of a substance or a mixture of compounds usually involves dissolving the organic compound in a suitable solvent, preferably one in which it is more soluble hot than cold. The solid can also be dissolved in a water-miscible solvent, which causes the substance to separate into crystals when water is added.

It is sometimes necessary to wash the crystals with water or other immiscible solvents that will extract impurities. For example, benzoate esters are washed with dilute sodium carbonate solutions to remove unreacted benzoic acids or to destroy excess benzoyl chlorides.

Solvent Choice

The proper choice of solvent is important in purifying derivatives by crystallization. Unfortunately the ideal solvent cannot be chosen strictly by theoretical considerations or rules of thumb. The literature on derivatives gives some information on the solvent to be used but seldom gives solubility data. However, the solubility can quickly be determined by using the gas-chromatographic technique (see Chap. 3). The purity can be determined on the first crop of crystals. If it is substantially purer than the original, the proper solvent has been used; if not, another must be found.

A few general rules will make selection of a solvent easier:

1. Since "like usually dissolves like," a solid will be most soluble in that liquid which it most resembles in structure. For example, solid esters are most soluble in ethyl acetate or other solvent esters (as well as in methanol or ethanol).

2. If possible, it is better to select that solvent in which the impure product is most soluble when hot but only slightly soluble when cold, thus retaining the impurities in the solvent.

3. Mixed solvents are useful when the impure compound is quite soluble in one of the cold solvents but only slightly soluble in the other cold solvent. For example, if the compound is very soluble in acetone,

it is dissolved in a small amount of this solvent; then a hydrocarbon is added slowly to force the crystallization.

4. In purification by crystallization, it is preferable for the compound being purified to be as soluble as possible and the impurities insoluble or much less soluble.   For example, a phenylurethan derivative of an alcohol dissolves in hot petroleum ether, but the impurity, diphenylurea, is insoluble.   Gas-chromatographic examination of the solvents before and after crystallization will establish whether the impurities are being removed.

5. Naturally the analyst avoids using a solvent that reacts with the compound being purified unless the impurities are removed in the reaction and the original compound thus purified is recovered.   For example, a variety of acidic substances can be purified by solution in aqueous sodium hydroxide and subsequent filtration to remove impurities; then treatment with mineral acids precipitates the original acidic substance.

6. Some compounds are so new that the literature does not give a suitable solvent.   Depending upon the nature of the compound, try:

a. Methanol, ethanol, or a mixture of the lower alcohols

b. Water, acetone, or a mixture of acetone and alcohol

c. Benzene, a mixture of benzene and toluene, petroleum ether, or a mixture of benzene and petroleum ether

d. Glacial acetic acid or aqueous acetic acid

Table 2-1 lists solvents and solvent pairs useful in the crystallization of various derivatives.   The ideal technique is to choose a solvent in which the compound to be purified is very soluble and then add to the solution a solvent in which the compound is insoluble or only slightly soluble.   Both solvents must be mutually miscible.   Heating the solutions may assist in the purification.   The solution should be examined before and after crystallization in the gas chromatograph to determine whether a degree of purification has been attained.

For substances that dissolve only with difficulty the more powerful solvents, such as dimethyl sulfoxide (**caution**: hazardous on skin!), tetrahydrofuran, nitro paraffins, etc., may be used alone or in combination.

Induction of crystallization can be either *chemical* or *physical*. Chemical induction involves addition of a *seed crystal*, another solvent, or a solid substance that will induce crystallization.   Physical induction involves warming or cooling, scratching, or local drying to produce induction sites for crystallization.

Amount of Solvent for Crystallization

The proper amount of solvent can be determined by mixing the derivative in various solvents, in accordance with procedures described in Chap. 3.   The solvent is shaken with the derivative and allowed to

TABLE 2-1   Solvents and Solvent Pairs for Derivative Crystallization[5]

Listed in Order of Decreasing Solvent Polarity

Solvent or solvent pair	Derivative to be crystallized
Water	Carboxylic acids, amides, and substituted amides
Methanol	Acetates, benzoates, 3,5-dinitrobenzoates, other esters, amides, *p*-toluidides, nitro and bromo derivatives, etc.
Methanol-water..........	Benzyl and *p*-nitrobenzyl esters, anilides, sulfonamides, picrates, semicarbazones, hydrazones, and substituted hydrazones
Ethanol................	Compounds similar to those listed with methanol and methanol-water mixtures; molecular complexes
Dioxane-water...........	Xanthylamides
Ethyl acetate	Quaternary ammonium salts, esters
Isopropyl ether	Quaternary ammonium salts
Acetone-alcohol..........	Osazones, bromo compounds, nitro compounds
Petroleum ether..........	Phenylurethans, α-naphthylurethans, etc.
Petroleum ether–benzene ..	*p*-Nitrophenylurethans, 3,5-dinitrophenylurethans
Benzene	Picrates, molecular complexes
Chloroform and carbon tetrachloride	Sulfonyl chlorides, acid chlorides, anhydrides

settle or separate, and then the solvent layer is injected into the gas chromatograph.   The relative areas of the peaks of the derivative to the solvent on the chromatogram will give the approximate degree of solubility in the solvent.   If the solid is not appreciably soluble, additional solvent may be added until the compound dissolves; or the mixture may be heated.   A solvent that dissolves a solid completely but shows no difference in solubility cold or hot is not useful for purifying derivatives.   Another solvent can be added to this mixture to make the solid separate without carrying down the impurities.   To be useful in purifying a derivative by differences in solubility in hot and cold solvent, the compound should be at least 5 times more soluble in the hot than in the cold solvent.

Occasionally, samples of unknowns will not crystallize until much of the solvent is removed.   This can usually be accomplished by warming the solution on a hot plate and blowing a stream of air or nitrogen over the surface until enough solvent is removed to induce crystallization. It is important not to carry the evaporation too far or impurities may be carried down with the purified crystals.

DETERMINING PURITY OF COMPONENT

The peak on the chromatogram should be examined carefully for evidence of contamination, using the technique of Crippen and Smith.[8] The compound should also be chromatographed on several columns with different degrees of polarity. Then derivatives prepared from the component may again be chromatographed on additional different columns (Fig. 2-1).

When it has been determined that the separated component is essentially one compound and the types of functional groups or elements present have been established, the identity of the compound must be ascertained exactly.

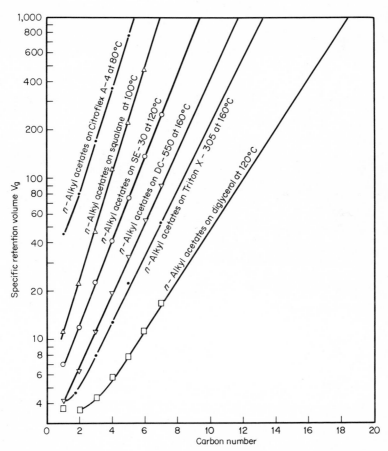

Fig. 2-1 Behavior of *n*-alkyl acetates on various columns and at various temperatures.[20]

DETERMINING IDENTITY OF COMPONENTS

If enough material is available or can be separated, the task becomes relatively easier. The melting point of a solid can be determined; for a liquid, the boiling point, refractive index, and other properties are measured. These properties can also be determined on micro quantities using the technique of Schneider[28] or by plotting retention data versus properties (see Chap. 4).

Using known similar compounds, gas-chromatographic retention values are determined on the known and unknown compounds.[20] With these data, a plot is made of the retention times versus the carbon number, as shown in Fig. 2-2. It is essential that a plot be made of the same type of compounds. For example, in the plot shown, a sample of polyol was separated from a mixture. The sample was composed of carbon, hydrogen, and oxygen with hydroxyl ($-OH$) groups present. Nuclear magnetic resonance (nmr) spectrometry showed that there were no terminal methyl groups. Three methylene ($-CH_2$) groups unassociated with hydroxyl and two hydroxyl groups attached to two terminal methylene groups were also present.

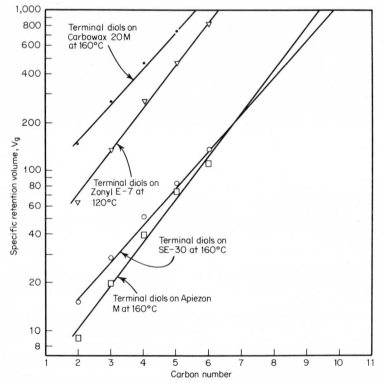

Fig. 2-2 Separation of terminal diols on various columns at different temperatures.[20]

A plot was made of 1,4-butanediol and 1,6-hexanediol.   The unknown sample had a retention time corresponding to 1,5-pentanediol.   Preparation of the acetates gave similar retention data.   Synthesis of pure 1,5-pentanediol verified all retention data and physical or chemical properties.   Infrared and nmr spectra corresponded to the unknown.

GROSS OBSERVATIONS

An experienced chemist can often find substantial clues to the probable identity from characteristic odors, colors, shapes of crystals, and other observable properties.   The functional-group class of a compound can be determined from careful study of the odor.   The color will also reveal the presence of a functional group.   Even a colorless compound may show the presence of a functional group when it is examined in the ultraviolet region.   The shapes of crystals are added clues to the type of groups that may be present.

After all the gross properties have been examined and recorded, they can be further expanded by optical examination of the crystals through a chemical or metallurgical microscope, which may reveal much about the probable identity of the substance.   For example, examination of a single crystal may show that it has optical properties upon heating or melting.   Examination of crystals with respect to their optical properties generally yields 10 to 12 characteristics that should assist in identifying the material.   With these gross and microscopic data, plus the gas-chromatographic and physical data, the probability of some other substance with the same set of properties is less than one in a million.

Caution: All compounds should be sniffed with extreme care.   Some are toxic or strongly noxious, and caution should be observed until more is known about the material.

FLAME-PYROLYSIS TESTING

Place a small quantity of the material (approximately 100 mg) in a crucible.   If the sample is a liquid, pass a flame over the surface to determine whether it will burn.   If the sample does not ignite, or if it is a solid, heat gently at first with the bunsen burner and then heat strongly.   Make a record of the following observations:

1. Flammability.
2. Flame properties (luminous, smoky, bluish, etc.).
3. Melting, subliming, evolution of vapors, change of color, etc., for a solid.
4. Odor.
5. If any residue is left in the crucible, perform the following tests:
   a. Does it fuse?

b. Does it dissolve in water?

c. Is it acidic or basic?

d. Does a gas evolve when dilute hydrochloric acid is added?

e. What is the nature of the metallic residue, if any?

Place a quantity of the material in a small test tube equipped with a stopper containing a short glass tube dipping into a solvent, such as chloroform. Heat the sample until it begins to pyrolyze and distills. Continue until the pyrolyzate bubbles into the solvent. Inject this pyrolyzate solution into the gas chromatograph. Much valuable information can be obtained from pyrolytic degradation of samples that are otherwise difficult to analyze. Many polymers can be degraded to their monomers for simpler identification.

Thus many clues to the identity of organic compounds can be obtained by flame or pyrolysis tests (see Table 2-2).

TABLE 2-2   Flame and Pyrolysis Tests

Flame Test		
Flame*	**Compound class**	
	Liquids	Solids (all melt and burn)
Sooty..............	Aromatic hydrocarbon	Aromatic hydrocarbon
Yellow.............	Aliphatic hydrocarbon	Aliphatic hydrocarbon
Whitish............	Ether	Solid ether
Clear blue..........	Alcohol, ketone, acid, or ester†	Solid alcohol, ketone, acid, or ester†

Pyrolysis Test		
Result‡	**Compound class**	
	Liquids	Solids (melt)
Volatilizes (no residue) or unchanged......	Undecomposable	Undecomposable solid
Volatilizes with some decomposition	Slightly unstable	Slightly unstable
Decomposes to gas and liquids............	Polymer	Solid polymer
Leaves residue.......	Salt of organic compound†,§	Salt of organic compound†,§

* List proceeds in order of less carbon and more oxygen.

† Carbohydrates are carbonized.

‡ List proceeds in order of decreasing stability.

§ May or may not be stable.

DETERMINING THE PRESENCE
OF WATER

Water is so universally present in our environment that most substances absorb some amounts unless precautions are taken to prevent it. Although in many tests, water does not interfere, in others, such as sodium fusion and detection of hydroxyl groups, the water must be removed. Water can be effectively removed from liquids by using molecular sieves or drying agents such as calcium sulfate or calcium oxide. **Caution**: do not use anhydrous magnesium perchlorate or similar powerful oxidizing agents, which may explode in contact with the organic material.

Water in liquids or in soluble solids can be detected by adding anhydrous copper sulfate. The copper sulfate changes from white to blue in contact with water. Water can also be detected on a gas chromatograph by using a highly nonpolar column such as a silicone oil or a hydrocarbon polymer column, e.g., Poropak Q.[13,31]

Solids usually can be freed of moisture by heating in an oven or vacuum. Even room-temperature storage over a drying agent in a desiccator may effectively remove most of the moisture.

TESTS FOR UNSATURATION

Unsaturated compounds vary in reactivity. Those which are easily oxidized or readily undergo addition reactions have active unsaturated bonds. Aqueous permanganate solutions oxidize these bonds quickly, with the disappearance of the purple color. Bromine (1 to 2%) in carbon tetrachloride is quickly decolorized without any evidence of hydrogen bromide evolution, indicating direct addition of bromine to the double bond.

Permanganate (Baeyer's) Reaction

Dissolve the unknown compound (about 0.1 g) to be tested in 2 ml of water, alcohol, or ether in a test tube and add a few drops of aqueous potassium permanganate (1%). Shake the solution. A rapid disappearance of the purple color is a good indication of a probable unsaturated group. Certain aldehydes, keto acids, or easily oxidizable hydroxy compounds destroy permanganate, though at a slower reaction rate. Check the oxidized product for change in retention time on the gas chromatograph.

Bromine in Carbon Tetrachloride

Dissolve the unknown compound (about 0.1 g) in carbon tetrachloride or glacial acetic acid. Add by dropper a 1 to 2% solution of bromine in carbon tetrachloride. If more than a few drops of bromine solution are required, the compound has unsaturated groups. If hydrogen bromide is evolved, as evidenced by acrid white fumes, substitution reactions rather than addition reactions are taking place. Phenols, aromatic amines, aldehydes, ketones, and other compounds with active methylene groups or active hydrogens may decolorize bromine with evolution of hydrogen bromide. Check the product for change in retention time on the gas chromatograph.

HYDROGENATION

The degree of hydrogenation can be determined by the reduction of the hydrogen pressure in a hydrogenation bomb. Some investigators have injected the sample plus hydrogen into a precolumn of activated nickel.[23] The peaks of the unsaturated compounds disappear from the gas chromatogram, and the peaks of the saturates increase (see Fig. 2-3).

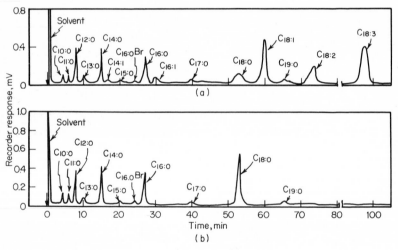

Fig. 2-3 Separation of fatty acid methyl esters on polyethylene glycol adipate column at 175°C (*a*) before and (*b*) after hydrogenation.[4]

FRIEDEL-CRAFTS REACTION (FOR AROMATIC SUBSTANCES)

Aromatic compounds react with anhydrous aluminum chloride in the presence of carbon tetrachloride or chloroform, producing colored complexes.

Dissolve 0.1 g of the unknown compound in 1 to 2 ml of dry carbon tetrachloride or chloroform in a test tube.   Introduce 0.2 g of anhydrous aluminum chloride, thus coating the walls of the tube above the liquid. Tilt the tube to wet the salt on the walls with the solution.   If aromatic compounds are present, the salt becomes colored, but the color in the solution is much slower to form.   Aliphatic or cyclic nonaromatic compounds do not produce colors as a rule.   Aromatic compounds and their halides produce colors ranging from yellow-orange to red; more complex aromatics produce blue to purple colors, and even more complex aromatics produce green colors (Table 2-3).

TABLE 2-3   Colors Produced in Friedel-Crafts Reactions

Compound	Color
Benzene and its homologs	Orange to red
Aryl halides	Orange to red
Naphthalene.	Blue
Biphenyl.	Purple
Phenanthrene.	Purple
Anthracene.	Green

The reaction solution may be examined by the gas chromatograph to verify that a reaction has taken place.   Even when aromatic compounds are not present, one can obtain a yellow color due to the presence of bromine or a purple or violet color due to the presence of iodine.   (See classification tests in Chap. 5 for further details.)

ELEMENTAL ANALYSIS

Carbon and Hydrogen

With the introduction of gas-chromatographic equipment for determination of carbon, hydrogen, and nitrogen (see Fig. 2-4), many laboratories routinely submit *all* samples to this analysis.   Most determinations take only 8 min to complete automatically.   A 10-mg sample is burned in the presence of a catalyst to carbon dioxide, water, and nitrogen.   The burned products are swept into the instrument by a stream of helium. The peak heights are measured and compared with the combusted standard.

An attachment for the gas-chromatographic carbon-hydrogen unit which can determine oxygen directly is due to be manufactured soon.

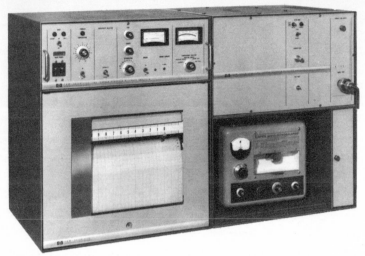

Fig. 2-4 Gas-chromatographic instrument for analysis of carbon, hydrogen, and nitrogen. (*Hewlett-Packard, Avondale, Pa.*)

Standard oxygen-determination equipment by the Unterzacher method is available from Arthur H. Thomas Co., Philadelphia.

Standard combustion-train equipment[14] for analysis of micro quantities of organic substances is available in most industrial laboratories, universities, or research institutions.

If combustion-train equipment is not available, the organic material can be ignited in the presence of copper oxide to form carbon dioxide and water. The products of combustion are dried by passing over anhydrous copper sulfate (turns blue if water or hydrogen is present) and are bubbled into a solution of barium hydroxide, in accordance with the method of Cheronis and Entrikin.[5] Reduction of the copper oxide to copper and the appearance of a white precipitate in the barium hydroxide solution indicate the presence of carbon. Hydrogen may or may not be indicated by the appearance of a blue color in the anhydrous copper sulfate layer, as some may be bound water.

Nitrogen

Nitrogen is detected in the gas-chromatographic combustion method as free nitrogen. In the Dumas method, it is combusted to free nitrogen and swept into the nitrometer with pure carbon dioxide, which is absorbed by a 50% potassium hydroxide solution, leaving the free nitrogen gas.

In qualitative procedures, the nitrogen is converted to the cyanide using sodium, magnesium–sodium carbonate, or zinc–sodium carbonate fusion procedures; or it is converted to ammonia using the calcium oxide fusion technique.

The sodium or magnesium–sodium carbonate fusions can be further analyzed for halogens, sulfur, or even phosphorus. In the zinc–sodium carbonate fusion, the sulfur is converted to insoluble zinc sulfide and remains on the filter paper when the solution is filtered. The insoluble zinc sulfide can be acidified and tested by means of lead acetate paper for the presence of sulfide.

Sodium Fusion. Extreme caution must be exercised in making sodium fusions, and the sample must be dry. Use only small quantities of sample and a small piece of sodium, cut under inert solvent and dried with filter paper. Handle the sodium metal with tongs or rubber gloves. Place 25 mg of sodium metal in a 4- to 6-in. flameproof test tube. Add about 10 mg of the solid to be tested or 1 drop of the unknown liquid. Allow to stand for approximately 2 min to determine whether the substance will react spontaneously with the sodium. Clamp the tube on a ring stand and heat with a small flame. Protect the face with a shield or goggles and the hands with rubber gloves. Heat the lower part of the tube until the sodium metal just melts. Remove the flame and add another portion of 10 to 20 mg of the unknown solid or 1 to 2 drops of the unknown liquid. Again heat the lower part of the tube strongly for 1 min and allow the tube to cool. When completely cool, add 1 ml of methanol to remove the excess free sodium. Then add 6 to 7 ml of water, heat to boiling, and filter. Test the filtrate for cyanide, sulfide, or halide ions. If cyanide ion is present, place 1 to 2 ml of filtrate in a test tube and add 0.1 to 0.15 g of solid ferrous sulfate. Heat the solution to boiling, add 1 drop of ferric chloride solution, and then add dilute sulfuric or hydrochloric acid until the solution is acid and iron hydrated oxides dissolve. Allow the solution to stand for a few minutes. A blue color or blue precipitate indicates the presence of nitrogen as cyanide. A clear solution or yellow tinge indicates the absence of nitrogen. A green or greenish-blue color indicates a poor or incomplete fusion. Repeat the fusion test.

The acidic solution may be injected into the gas chromatograph on a 7 ft $\times$ 0.3 in. 20% Carbowax 1500 column under conditions discussed in the article by Leibrand.[17] A peak with identical retention, as illustrated by Leibrand, indicates nitrogen as hydrogen cyanide.

If *sulfur is present*, it may obscure the nitrogen test. For this determination use 1 ml of the solution; then add 0.1 to 0.15 g of ferrous sulfate and sodium hydroxide until the solution is distinctly alkaline. Heat to the boiling point and filter. Add a few drops of ferric chloride to the acidified solution and allow it to stand for a few minutes. Nitrogen will develop a blue color or blue precipitate.

Magnesium–Sodium Carbonate Fusion. In a small flameproof test tube add 50 mg each of dry magnesium powder and anhydrous sodium carbonate; then mix in 50 mg of the unknown sample. Clamp the tube on a ring stand and add 100 mg of the magnesium–sodium carbonate mixture on top of the sample. Heat the mixture at the area farthest from the sample. When the sample begins to burn, heat the lower end of the tube until it is red hot. Cool and add 5 ml of water, boil, and filter. Test filtrate for nitrogen (as cyanide) as described under sodium fusion.

Zinc–Sodium Carbonate Fusion. In a 50×6 mm fusion tube, put 0.15 g of a 50/50 mixture of zinc dust and anhydrous sodium carbonate. Mix in 30 to 50 mg of the solid sample or 1 to 2 drops of the liquid sample and add another 0.15 g of the zinc–sodium carbonate mixture. Clamp the tube on a ring stand at an angle and heat above the mixture, gradually approaching the sample until the bottom is red hot. Drop the hot tube into a larger test tube filled with 5 to 8 ml of distilled water. Break up the shattered tube, boil, and filter. The filtrate should be colorless. Test for cyanide as described in the section under sodium fusion. Acidify under a hood and inject the solution into the gas chromatograph to detect hydrogen cyanide, HCN, as described by Leibrand.[17]

Calcium Oxide Fusion. Mix into 0.2 to 0.3 g of powdered soda lime 50 to 100 mg of the compound to be analyzed. Place in a small evaporating dish and cover with a watch glass to the lower side of which has been attached a strip of moist red litmus paper. Heat slowly and observe the paper. A change to blue indicates nitrogen present as ammonia. The basic solution or the vapors may be injected into the gas chromatograph using a 5 ft $\times$ $\frac{1}{4}$ in. Carbowax 400 column treated with potassium hydroxide. A peak like that illustrated by Leibrand[17] with an identical retention time indicates ammonia is present. Amines have different retention times.

The cooled solution in the evaporating dish may be tested for sulfide ion.

Test for sulfur by placing about 1 ml of the filtrate from the sodium fusion in a test tube acidified with dilute acetic acid. Add 1 to 2 drops of lead acetate solution. The presence of sulfur is indicated by a brown or black precipitate or color. For a confirmatory test, add 1 drop of a 0.1% sodium nitroprusside solution to 0.5 ml of the solution to be checked. A deep red color confirms the presence of sulfide.

Injecting the acidified filtrate onto a 6 ft $\times$ $\frac{1}{4}$ in. 30% Triton X-305 column under the programmed conditions (25 to 75°C) described by Leibrand[17] will give a hydrogen sulfide peak. A black spot on a strip of moist lead acetate paper as the gas emerges from the instrument confirms the presence of hydrogen sulfide.

In the zinc–sodium carbonate fusion, any sulfur present will be in the precipitate rather than the filtrate.   The filter paper may be acidified in a small beaker covered with a watch glass to the underside of which is attached a strip of lead acetate paper.   The paper turns black if sulfur is present.   The solution may also be injected into the gas chromatograph as described above.

For the *halogen tests* the filtrate from the sodium fusion is acidified with dilute nitric acid and boiled under the hood to remove any cyanide and sulfide that may be present.   Silver nitrate solution is added to a portion of the solution: a white precipitate indicates chloride, a yellowish-white precipitate indicates bromide, and yellow precipitate indicates iodide.   For confirmatory tests, acidify another portion of the filtrate with dilute sulfuric acid and add a layer of chloroform plus a few drops of chlorine water or 4 drops of hydrogen peroxide solution (3%).   The chloroform layer is colorless if the chloride ion is present, brown or tan if the bromide is present, and violet if the iodide is present.   The free halogens or hydrogen halides in the chloroform layer may be injected into the gas chromatograph as shown in Fig. 2-5.   If hydrogen halides or halogens are injected frequently, Teflon or gold-coated detectors should be used.   Flame ionization detectors should not be used.

The Beilstein halogen test depends upon the effect of the vapors of the copper halide ions on the color of the flame.   Heat a clean copper wire in a colorless flame until it no longer produces a color.   Touch the wire to the compound to be tested and reheat in the colorless portion of the flame.   A blue-green flame indicates the presence of halogens. This test can be used as a halide detector on the chromatograph.   When the effluent peaks emerge from the instrument, they can be directed against the hot copper wire.   The component is a halide if the flame at the effluent turns blue-green.

If a white residue remains after ignition, test it for metallic ions. The residue may be a metal oxide or a carbonate.   Ignite 0.3 to 0.5 g of the substance in a crucible for 15 to 20 min at a dull red heat.   Cool, moisten the residue with a few drops of concentrated sulfuric acid or aqua regia, and heat until the residue dissolves.   Silica and boron oxides do not dissolve except in strong alkalies.   Mercury is volatilized during ignition, but it usually is detected by oxidizing 0.2 g of the substance with 10 ml of concentrated potassium chlorate until colorless. The mercury ion is then reduced with a piece of copper wire.   An amalgam on the wire indicates the presence of mercury.   Tests for phosphorus and arsenic are performed on the compound fused with a mixture of potassium nitrate and anhydrous sodium carbonate.

Further tests for other metallic ions are performed by the usual qualitative inorganic scheme.[1]   If emission spectrographic equipment is available or even atomic absorption units, the residue can be rapidly

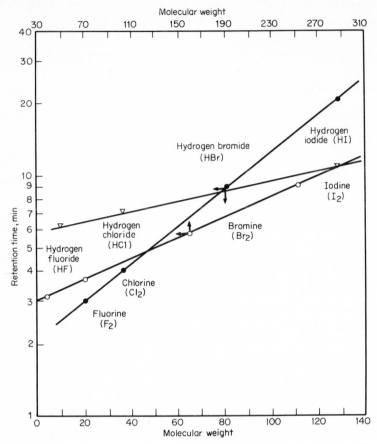

Fig. 2-5   Elution of hydrogen halides and halogens on Kel F 40 oil on Teflon (30 to 60 mesh) at 48°C (argon: 16 ml/min) and Poropak QS at 80°C (1.5 ft × ¼ in.).[3,17]

examined for metallic elements.   At Northern Kentucky State College, the author combined the atomic absorption unit with the gas chromatograph as a metal detector.[7]

Other investigators[9] have converted metallic ions into stable organic complexes.   Many of these are volatile enough to be chromatographed and can be measured fairly accurately (see Fig. 2-6).   However, since not all elements are volatile enough to be chromatographed (Fig. 2-6), no complete scheme can be offered at present for the examination of all the inorganic elements.   It is highly probable that eventually it will be possible to convert all the elements into volatile metalloorganic complexes that can be examined on the instrument.   To make the scheme as foolproof as possible, one complex should be capable of converting all

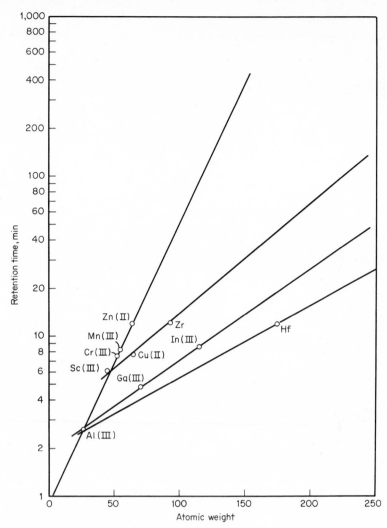

Fig. 2-6   Separation of metal trifluoroacetylacetonates on 4 ft × 4 mm glass tubing with DC-710 (0.5%) silicone oil on micromesh beads at 125°C (helium: 83 ml/min).[26]

inorganic elements into volatile complexes, which should be examined on a single column for separation of all known elements according to molecular weight (for further information see Ref. 26).

GAS-CHROMATOGRAPHIC TECHNIQUE

In a gas-chromatographic technique described by Martin and Synge[25] in 1941 a liquid supported on a solid was used as a fixed phase. These authors used column chromatographs with mobile solvents as well as paper chromatography and suggested that it might be possible to utilize a liquid supported on a finely powdered solid to separate gases or volatile organics in the vapor phase.   The first practical gas chromatograph was developed a little over a decade later.   The developments in equipment were extremely rapid and are continuing apace.

There are many good texts on the theory of gas chromatography,[9,18,27] to which the reader is referred.   No attempt is made to discuss the theory of gas chromatography here, but a few terms will be defined (see Fig. 2-7).

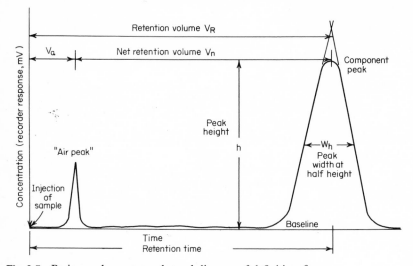

Fig. 2-7   Basic gas chromatography and diagram of definitions.[2]

Retention time t_R (or elution time) is the elapsed time necessary for a component to be eluted from a column under a given set of conditions. From the point of injection, the retention time is

$$t_R = \frac{(1+K)L}{u}$$

where K = column-capacity coefficient
L = length of column
u = true linear velocity in column

Relative retention time t_{RR} is the ratio of the retention time of the compound to that of a reference standard

$$t_{RR} = \frac{t_R \,(\text{compound})}{t_R \,(\text{standard})}$$

It is a convenient reference for relating the retention time of the same compound on various columns versus a known standard compound.

Retention volume V_R, or true retention volume, is defined as the volume of gas passing between injection of the sample and the time for the peak of the component to reach its maximum, measured at the average temperature and pressure of the column. The retention time is related to the retention volume as follows:

$$t_R = \frac{V_R}{f_{av}}$$

where f_{av} is the average flow rate through the column. The *true retention volume* can be obtained from the experimentally obtained retention volume V_E according to the equation

$$V_R = \frac{V_E(TP_r)}{T_r P_{av}}$$

where T = temperature of column
T_r = temperature measured externally in laboratory
P_r = pressure measured externally in laboratory
P_{av} = average pressure of column

The *net retention volume* V_R' is the true retention volume less the dead volume of the column V_a. This is the retention volume of air which is equal to the volume of gas in the column when measured at the average column pressure:

$$V_R' = V_R - V_a$$

The *specific retention volume*[19] V_g corresponds to the volume of gas, measured at column-outlet pressure and corrected to 0°C, required to move one-half of the solute through a theoretical column containing 1 g of liquid phase. This theoretical column has no pressure drop or dead volume[20]

$$V_g = \frac{V_n \times 273}{W \times T}$$

where V_n = net retention volume
W = weight of liquid phase in column, g
T = temperature at which gas flow was measured, K

This equation uses the correction factor of Martin and James.[24]

Band broadening is the tendency of a component moving down a column to spread out.

Column efficiency is the ratio of band narrowness to its retention volume. This relationship serves as a measure of the efficiency of the column. Column efficiency = V_R/w, where V_R is the retention volume and w is the width at half height.

Peak resolution R is

$$R = \frac{t_{R2} - t_{R1}}{w_b} = \frac{t}{w_b}$$

where $t_{R2} - t_{R1}$ is the separation time between peaks and w_b is the width at the base.

Column capacity is the ability of any given column to hold a definite volume of a component.   More than this volume overloads the column and distorts the peak.   The peak becomes skewed and if the overload is great enough, may obscure other components.   Overload is desirable only when one is looking for low levels of impurities and when it does not obscure their detection.

HYDROCARBON INDEX NUMBER

The hydrocarbon index number I_x was defined and calculated by the method of Kovats:[16]

$$I_x = 200 \frac{\log V_n x - \log V_n z}{\log V_n(z + 2) - \log V_n z} + 100z$$

where V_n = retention volume
z = number of carbons in first normal hydrocarbon standard
x = compound whose retention is between that of first and second normal hydrocarbons

It is used for calculating index numbers of hydrocarbons and hydrocarbonlike compounds.   It has been proposed for use in the petroleum industry and has recently been expanded to include nonhydrocarbon compounds. (For more detailed definitions and relationships, see Ref. 1.)

If the degree of separation desired cannot be obtained with the packed columns, capillary columns or support-coated open tabular columns (SCOT) may be used.

Figure 2-8 shows the variation of retention time with temperature for *n*-hexane, cyclohexane, and benzene.   At increasing temperatures, poorer efficiency of peak resolution results.

LIQUID PHASES AND GAS-CHROMATOGRAPHIC COLUMNS

All columns prepared by McReynolds[22] were made using Celite 545 (diatomaceous silica).   The fines were removed by water washing, and coarser material was collected on a suction filter, dried, and screened through 60- to 70-mesh standard screens.

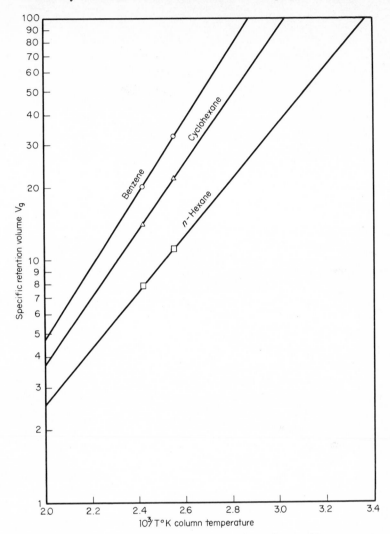

Fig. 2-8   Hydrocarbons: variation of specific retention volume with temperature on dioctyl phthalate column.[21]

In preparing the packing, all but the polyester columns were made by adding 0.5 percent of a wetting agent (Poly Tergent J-300) to the liquid phase to reduce the tailing of polar compounds and increase column efficiency.   The liquid phase (20 to 25 percent) was added to 75 to 80 percent of the dry packing material slurried in acetone or methylene chloride.   The mass was stirred constantly until almost dried and then spread on a glass plate until free of solvent.

The column packing was pressure-packed into ³⁄₁₆-in.-OD tubing of 4- and 12-ft lengths.   Columns of 25-ft lengths were made of highly polar and highly nonpolar liquid phases.   Compounds having long retention times were run on the shorter columns, and compounds having short retention times were run on the longer columns.   McReynolds[20] reported on 75 packing materials used in his studies.

The sample size injected into the instrument depends on whether the material is dissolved in a solvent and whether one is looking for major components or trace impurities.   Usually 0.5 μl is used for high-purity materials and 2.0 μl for compounds of average purity; 10 μl or more is used for determining trace impurities.

PRACTICAL GAS CHROMATOGRAPHY

Column Preparation (Inert-Support Treatment)

Gas-chromatographic columns are composed of fine particles in an inert support, from 40- to 150-mesh or finer, coated with a liquid phase. It is essential to reduce active sites on which organic components might be adsorbed by washing the inert support with acids, bases, or various solvents.   Even with these treatments, highly polar substances, such as alcohols, ketones, or water, give tailing of the peak, which indicates a considerable degree of adsorption.   Further treatment with trimethyl-silylating agents, such as hexamethyldisilazane, trimethylchlorosilane, and pyridine, involves injecting at least 20 μl into the column while holding the column temperature at approximately 250°C; this reduces or eliminates tailing (see Fig. 2-9).

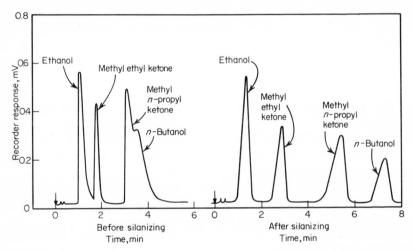

Fig. 2-9   Effect of silanizing Chromosorb P (80/100 mesh) coated with dinonyl phthalate (20%).[11]

Coating Inert Support. The inert supports are coated with the liquid phase by dissolving it in a solvent, slurrying with the inert support, and evaporating the solvent while stirring the mass to ensure uniformity. Some investigators have poured the solution of the liquid phase under vacuum through the inert support. A certain portion of the liquid phase remains on the inert support after drying (usually about 2 percent). The actual percentage can be estimated by evaporating the solvent solution passing through the inert support and weighing the residue. After the inert support is well coated and most of the solvent is evaporated on a steam bath in a hood, the nearly dry residue is spread on a nonabsorbent plate until thoroughly dry.

Packing Column. A metal column of suitable length and outside diameter is selected. A glass-wool plug (preferably silylated) is inserted in the lower end of the tube. A plastic funnel is fastened to the top of the column by means of a piece of rubber tubing. By gentle tapping, the dry packing is fed into the column until full, allowing space for another plug of glass wool in the upper end.

If a pressure filler is available (Chemical Research Services, Inc., Addison, Ill.) the column is filled under 15 to 20 lb of air or nitrogen pressure until no more material enters the column with continued tapping or vibration.

Swagelock fittings are attached to both ends and the column bent to fit the instrument and connected to it. Leaks should be checked for with a soap solution.

Conditioning Column. The gas flow is adjusted through the column and heated to the recommended temperature for conditioning. The heating and gas flow are continued until a stable base line is obtained. When the column is adjusted to the operating temperature, it is ready for use.

Injection of Samples

Samples should be liquids or solids in solution. With the solid-injection system available from Hewlett-Packard, Avondale, Pa., certain volatile or decomposable solids can be injected directly into the gas chromatograph. Liquid samples are injected into the instrument with microsyringes, available from numerous manufacturers such as the Hamilton Co., Whittier, Calif.

The syringe is overfilled above the desired mark to ensure that no air bubble is trapped in the nontransparent part of the metallic needle and then readjusted to the mark. The finger must be held alongside the plunger so that the pressure does not blow the plunger out of the syringe. The point of the needle must be firmly pressed through the center of the septum in the injection port and the plunger quickly depressed to obtain a "plug" injection. Too slow an injection produces broad peaks. The needle is withdrawn immediately after the injection to prevent it from overheating.

In filling syringes with removable needles, care must be taken that the "hub" of the needle is filled with the sample. This type of syringe is useful for viscous liquids. It is filled without the needle in place; then the needle hub is filled and attached to the syringe. Injection is done in the usual manner to ensure a plug injection. If the pressure of the injection tends to blow off the needle, the needle can be held on firmly by placing the fingernails behind the hub ridges during the injection (being careful not to let the fingers touch the hot injection port).

One more note of caution. The plunger of the 10-μl syringe is very delicate and easily bent. When the plunger is pushed, it must be held straight, as a sidewise pressure may bend the plunger. Similarly, unsheathed needles of fine bore are easily bent, and extreme care is necessary in pushing the needle through the silicone septum to avoid bending the tip or the body of the needle.

Septums should be changed frequently, especially when using larger-bore or sheathed needles. Many injections tear the septums, allowing the sample to leak or "spit" out and excessive air to leak in. Often when good peaks cannot be obtained, a change of septum enables the instrument to perform satisfactorily. Hamilton Co.[12] has a multilayer septum that is said to last many times longer than standard types.

GENERAL DISCUSSION

If the unknown sample is in the form of coarse crystals or lumps, pulverize it to increase its ability to go into solution. Similarly, if the unknown solid (or liquid) seems to be insoluble in water or ether, heat the mixture carefully on a *hot* plate (**Caution**: Ether is very flammable! Use in ventilated hood and avoid flames). When the sample goes into solution, cool it to room temperature and shake vigorously to prevent supersaturation. Add a seed crystal of the unknown to allow excess to crystallize.

On the basis of the solubility of organic compounds in selected solvents, chemists have classified compounds for identification and characterization.[5,15,29,30] The introduction of gas chromatography has enabled the chemist to determine the degree of solubility more accurately. A slight solubility in water, for example, may indicate presence of polar groups such as hydroxyl in a fatty alcohol. Gas chromatography can indicate small solubilities (below 3 percent).

The schemes to be used in this book combine the classification by solubility with determination of the elements present in the compound and gas-chromatographic examination. This technique further reduces the possible chemical classifications which require functional tests to identify the compound.

To demonstrate the solubility-classification relationships for identifying an unknown organic compound, the following steps should be followed:

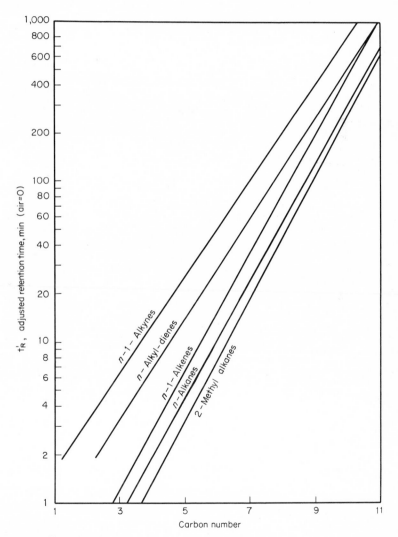

Fig. 2-10 Plot of various homologous series of hydrocarbons vs. carbon number on dimethylsulfolane at 50°C.[10]

1. Determine the elements present.   Usually, one can assume that carbon and hydrogen are present in most organic unknowns.   If there is a residue on ignition of the compound, analyze it for metallic or nonmetallic elements.

2. Using the classifications from the elements present and the solubilities as determined, assign your unknown to the pertinent group. In order to identify the compound, it is *not* necessary to test the entire organic field, but only a few functional groups.   Behavior on various

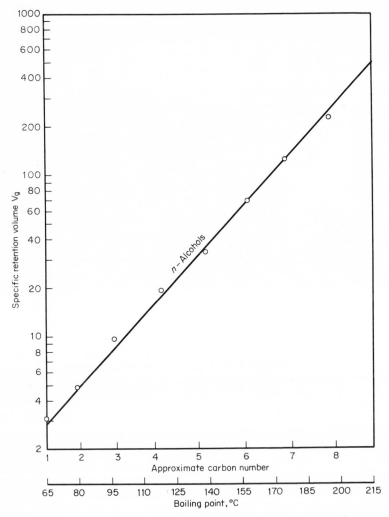

Fig. 2-11 Specific retention volume of *n*-alkyl alcohols on SE-30 (silicone gum rubber) at 120°C vs. boiling point (estimate of carbon number).[20]

gas-chromatographic columns may also help determine the probable identity of the compound.

3. Use the chemical reactions in which the functional groups present will take part to identify the unknown compound. Reactions can be performed on reactive gas-chromatographic columns or pre-columns. Retention times before and after reaction may be compared. These functional-group tests are listed in Chap. 5. Thus, it is possible to classify the compound as belonging to a specific chemical class.

4. Relying on the retention data for the unknown compound and for higher- and lower-boiling (or melting) compounds, plot a homologous-series curve of log retention time versus carbon number (see Fig. 2-10). Determine the various physical properties of the purified compound. If the property is steadily increasing, it can be plotted versus the retention time or volume. The number of carbon atoms in the compound can be predicted (Fig. 2-11). Using both chemical and physical properties, it should now be possible to narrow the probabilities of the unknown to a few compounds. The elemental composition, added to the chemical and physical properties, should produce even further restrictions on identity.

5. To verify the final identity, make one or more derivatives and determine their properties on various columns. If the data fall within the predicted retention time for known similar derivatives of homologs which are higher and lower, the unknown has been effectively bracketed with knowns. Determination of the physical properties should verify the identification, e.g., infrared curves, ultraviolet absorptions, nmr data, or mass spectral curves, together with density, refractive index, etc.

REFERENCES

1. Arthur, P., and O. M. Smith: *Semi-micro Qualitative Analysis*, pp. 177, 244, McGraw-Hill Book Company, New York, 1952.
2. ASTM: *Proposed Recommended Practice for Gas Chromatographic Standards*, pt. 30, p. 875, Philadelphia, 1967.
3. Benner, R. S.: Aquaphase Laboratories, Inc., Adrian, Mich., private communication, 1967.
4. Burchfield, H. P., and E. E. Storrs: *Biochemical Applications of Gas Chromatography*, p. 541, Academic Press Inc., New York, 1962.
5. Cheronis, N. D., and J. B. Entrikin: *Semi-micro Qualitative Organic Analysis*, p. 13, Thomas Y. Crowell Company, New York, 1947.
6. *Ibid.*, pp. 88–91.
7. Crippen, R. C.: Unpublished papers.
8. Crippen, R. C., and C. E. Smith: *J. Gas Chromatogr.*, **6**: 37–42 (1965).
9. Dal Nogare, S., and R. S. Juvet, Jr.: *Gas-Liquid Chromatography*, p. 390, Interscience Publishers, New York, 1962.
10. *Ibid.*, p. 417.

11. *Ibid.*, p. 144.
12. Hamilton Co., Whittier, Calif.
13. Hollis, O. L., and W. V. Hayes: *J. Gas Chromatogr.*, **4**: 235–239 (1966).
14. Ingraham, G.: *Methods of Organic Elementary Micro-analysis*, chap. 1, Reinhold Publishing Corporation, New York, 1962.
15. Kamm, O., *Qualitative Organic Analysis*, 2d ed., John Wiley & Sons, Inc., New York, 1932.
16. Kovats, E.: *Z. Anal. Chem.*, **181**: 351 (1961).
17. Leibrand, R. J., Jr.: *Gas Chromatogr.*, **5**: 518–524 (1967).
18. Littlewood, A. B.: *Gas Chromatography*, Academic Press Inc., New York, 1962.
19. Littlewood, A. B., C. S. G. Phillips, and D. T. Price: *J. Chem. Soc. (Lond.)*, **1955**: 1480.
20. McReynolds, L. O.: *Gas Chromatographic Retention Data*, p. 144, Preston Technical Abstracts Company, Evanston, Ill., 1966.
21. *Ibid.*, p. 73.
22. *Ibid.*, pp. 3–20.
23. Markley, K. S.: *Fatty Acids*, 2d ed., pt. 3, p. 2184, John Wiley & Sons, Inc., New York, 1964.
24. Martin, A. J. P., and A. T. James: *Biochem. J.*, **50**: 679 (1952).
25. *Ibid.*, **35**: 1358 (1941).
26. Moshier, R. W., and R. E. Sievers: *Gas Chromatography of Metal Chelates*, p. 121, Pergamon Press, New York, 1965.
27. Purnell, H.: *Gas Chromatography*, John Wiley & Sons, Inc., New York, 1962.
28. Schneider, F. L.: *Qualitative Organic Micro-analysis*, p. 108, Academic Press Inc., New York, 1964.
29. Shriner, R. C., R. C. Fuson, and D. Y. Curtin: *Systematic Identification of Organic Compounds*, 4th ed., John Wiley & Sons, Inc., New York, 1959.
30. Standinger, H.: *Auleitung zur organische qualitative Analysis*, J. Springer, Berlin, 1925.
31. Waters Associates: *Poropak Bulletin*, Framingham, Mass., 1967.

CHAPTER THREE

Solubility Studies

Gas-chromatographic procedures for studying the solubility behavior of organic compounds have greatly improved the sensitivity of the observations. For example, the slight solubility that is not detected by standard solubility techniques may indicate functional groups that are masked by the larger portion of the insoluble molecules. *n*-Butanol is considered soluble in water, whereas *n*-pentanol is considered insoluble.[46] For example 1 part of *n*-butanol is soluble in 15 parts of water; less than 3 percent *n*-pentanol is soluble, and *n*-hexanol is still less soluble. After thorough shaking and standing, these solubilities can be determined by injecting the water portion of the alcohol-water mixture into the instrument. The ratios of the peak areas (water/alcohol) will give a good estimate of the solubilities of the alcohols in the water at the measured temperatures. Similarly, the solubility of water in the alcohols can be measured by injection of the alcohol layers (see Fig. 3-1*a* and *b*).

Identical measurements can be made on the solubility in ether, acid, and base solutions. The observations serve also to reveal various clues to the nature of the molecule:

1. The presence of a functional group may be revealed. For

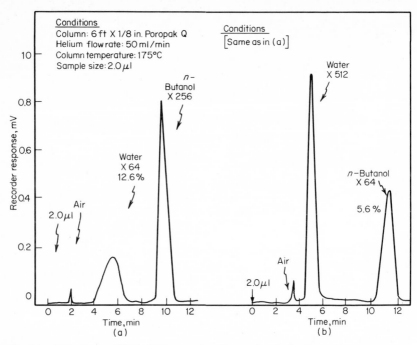

Fig. 3-1   Solubility at 25°C of (*a*) water in *n*-butanol and (*b*) *n*-butanol in water.

example, partial solubility in water may indicate that a functional group may be present.   Methyl ethyl ketone is slightly soluble in water, as is diethyl ether.

2. Solubility in specific solvents may give more specific data concerning the identity of the functional group.   For example, salicylic acid is insoluble in water (a polar solvent); this acid is converted to a water-soluble compound by reaction with dilute sodium hydroxide (5%) or dilute sodium bicarbonate solution (5%), which indicates that an acidic functional group is present.

3. Certain observations regarding molecular weight can be made. For example, if the substance is a monofunctional compound with less than five carbon atoms, the compound is water-soluble; compounds with more than five carbon atoms are relatively insoluble.   Fortunately, gas chromatography reveals the exact solubility of the substance in water.   The solubility of the alcohols in water is shown in Fig. 3-2. The curve indicates that 1-hexanol is soluble to the extent of 0.6 g/100 ml and 1-pentanol is soluble to the extent of 2.0 g/100 ml at 25°C. Solubilities below five carbon atoms increase, whereas solubilities above five carbon atoms decrease.

The solubility classes of organic compounds have been summarized in

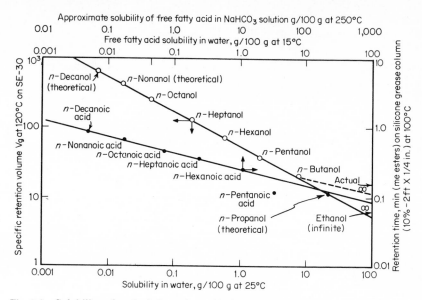

Fig. 3-2 Solubility of *n*-alcohols and *n*-acids in water and NaHCO₃ solution at 25 and 15°C vs. their specific retention volume on SE-30 at 120°C.[25]

Fig. 3-3. Other investigators have used various solubility schemes,[20,32] but the illustrated procedure seems the most versatile.

Compounds are divided into two main classes by their solubility in water (W_1 and W_2). The exact extent can be established by the gas chromatograph. Each of these is further subdivided into additional classes by their solubility in various solvents. As shown in Fig. 3-3, the water-soluble class is further subdivided by the solubility in ether, those soluble (E_1) and those insoluble (E_2). For purposes of the figure, the authors have arbitrarily called a compound soluble if it dissolves to the extent of 3 g or more per 100 ml of solvent and insoluble if it dissolves to an extent less than 3 g/100 ml. Gas chromatography can establish the *exact* extent of solubility and assist in determining whether functional groups are present.

Solubility in dilute acid or base represents a different problem. Instead of determining whether the compound is soluble to the extent of 3 percent, one finds whether it is much more or less soluble in the dilute acid or dilute base than in water alone. A definite increase in solubility represents a more positive indication of the presence of basic or acidic functional groups. Here again, the exact extent can be determined by acidification or alkalinization of the solution and subsequent injection into the instrument.

Acidic substances can be detected by their solubility in dilute sodium

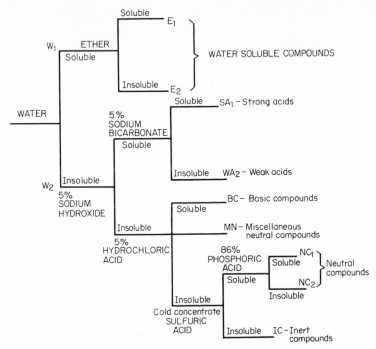

Fig. 3-3 Separation of organic compounds into various solubility classes according to Shriner, Fuson, and Curtin.[37]

hydroxide solution (5%). Accordingly, Shriner, Fuson and Curtin[41] have divided this class into strong and weak acids (classes SA_1 and WA_2), using a differentiating solvent, a mildly basic sodium bicarbonate solution (5%). Strong acids dissolve in this solvent; weak acids do not. Probably an intermediate class could be established from injection of the acidified sodium bicarbonate solution into the gas chromatograph. Partial solubility of one intermediate acid may be demonstrated (see Fig. 3-2).

Basic substances (class BC) are revealed by their solubility in weak hydrochloric acid solution (5%). The authors[36] in their classification have made no attempt to further divide this class into strong and weak bases. Some investigators have attempted to classify basic substances gas-chromatographically by their behavior on polar columns.[3,8]

Certain compounds that are essentially neutral in strong acid solutions (aqueous) behave like basic substances in the presence of more acidic solvents such as syrupy phosphoric acid and cold concentrated sulfuric acid. These compounds may contain sulfur or nitrogen or have an atom with an unshared pair of electrons. As expected, these compounds dissolve in the strongly acid solvent.   Substances which contain nitrogen or sulfur and which are neutral to acid or base are placed in class MN.

Neutral compounds in water having oxygen in any form act as fairly strong bases in cold concentrated sulfuric acid.   Similarly, any reactivity with sulfuric acid or any solubility in this solvent indicates that the unknown compound may contain oxygen or may have such hydrocarbon bonds as olefins or aromatic rings.   These compounds have been classified as neutral compounds (NC).   This class has been further subdivided by its reaction with 85% phosphoric acid: class NC_1, in which they are soluble, and class NC_2, in which they are insoluble. Inert compounds that are too unreactive or so weakly basic that they are essentially insoluble in cold concentrated sulfuric acid are placed in class IC (inert).   As with other classes, even slight solubilities can be detected by neutralizing the acid with which the compound is shaken, using the base, and separating the unknown from the mass of salts.

The simple study of the solubility of an unknown in water provides no evidence of acidic or basic groups on the molecule.   Routinely, the analyst should test the water solution for acidity or basicity.   He should also test a substance which is soluble in dilute hydrochloric acid solution (5%) for solubility in dilute sodium hydroxide solution (5%) since both functional groups may be present in the unknown molecule.

SOLUBILITY PROCEDURES

Macro Techniques

Using a long narrow graduated cylinder of 10 ml capacity, add 3 ml of the solvent to be used plus 0.2 ml of the unknown liquid or 100 mg of solid.   Cover with a clean tight-fitting stopper, mark the liquid level, and shake vigorously.   Allow the layers to separate and note whether the interface has increased, decreased, or remained unchanged. Record all observations.   If a solid, note whether the solid has vanished, decreased, or remained unchanged.   If necessary, remove solid, dry, and weigh.

Using a long needle, remove 2 to 10 μl of the upper layer (if any) and inject into a gas chromatograph.   Measure the areas of the solvent and unknown on the chromatogram.   The ratios of the areas of the peaks will give a relative indication of the solubility in the upper layer.   Repeat with the lower layer.   The solubility of n-butanol in water (lower layer) and water in n-butanol (upper layer) was shown in Fig. 3-1a and b.

The chromatographic technique may be repeated with water, ether, and hydrochloric acid solubilities, but usually the reaction with sodium hydroxide solution (5%) and sodium bicarbonate solution (5%) will not release volatile components until neutralized.   Even after neutralization, if the compound has been sulfonated, the sulfonate or phosphonate must be converted into a stable volatile compound before it can be observed.

Semimicro Technique

If quantities of the sample are limited, use 30 mg of the solid sample in 1 ml of solvent or 1 drop of liquid in 15 drops of the solvent in a narrow 3-in. test tube.   Stopper well and shake vigorously for 2 min before observing whether the compound has dissolved.

Solvents that can be used for the solubility determination should be tested in the following order:
1. Water
2. Ether
3. Aqueous hydrochloric acid (10%)
4. Aqueous sodium hydroxide (10%)
5. Aqueous sodium bicarbonate (10%)
6. Concentrated sulfuric acid
7. Syrupy phosphoric acid

Determine all solubilities at ambient temperature.   To speed up solution of slowly soluble substances, warm the mixture slightly, cooling the solution to room temperature again before deciding on the extent of solubility.

To test the compound in dilute sodium hydroxide, sodium bicarbonate, or hydrochloric acid solutions, shake the mixture of the sample and the test solution vigorously, separate by decanting (or filtering if necessary) the water solution from any undissolved solid (or liquid unknown), and carefully neutralize with acid or base.   Examine the solution for any separation of the original unknown material.   The appearance of slight cloudiness upon neutralization is a positive test of some solubility. Inject the neutralized solution into the instrument to determine the exact extent of solubility. If the compound is nonvolatile or decomposes, a derivative may be prepared to make it volatile or stable.

To verify your guesstimate, inject the lower layer into the instrument. The area ratios will give a fairly close estimate of the relative solubility. If there are two layers, inject samples of the upper layer.   If no peak occurs with the alkaline solutions, neutralize and inject a portion into the instrument.   Repeat the above prodecure with base if no peak is obtained for the acid solutions (see Fig. 3-4).

The beginning chemist may try all solvents, but it is usually not necessary to determine the solubility in every solvent listed.   If the scheme shown by Cheronis and Entrikin[7] is used, a compound that is

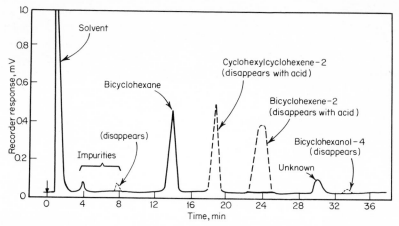

Fig. 3-4 Separation of cyclohexane isomers and impurities before and after sulfuric acid treatment (analyzed on 30% diethylene glycol succinate; 10 ft × ¼ in. at 145°C).

soluble in water may be tried in ether only, using no other solvent. A water-insoluble compound that is soluble in dilute hydrochloric acid may be tried in sodium hydroxide (10%) solution for amphoteric properties.

Micro Technique

Where samples of material are extremely limited, e.g., fractions collected from preparative gas-chromatographic instruments, the substance may be introduced into the solvent in a cell under the microscope.   If the material is crystalline, a microcrystal may be introduced on the tip of a spatula or micro stirring rod.   If the material is a liquid, a microdrop may be introduced by means of a microsyringe.   The droplet or crystal can be watched for decreases in size as the solvent around it is stirred. Thus, many solvents can be tried with very small amounts of sample. Gas chromatography can verify solubility or the lack of it.

Solids. If the unknown is a solid, it should be finely ground to a powder to increase its surface area, making it easier to dissolve.   When the solid does not appear to dissolve in either water or ether, it may be warmed in these solvents to enhance solution.   It should be cooled to room temperature again and shaken to prevent supersaturation.   If the material still does not crystallize on cooling, seed it with another small crystal of the solid and shake vigorously.   Examine the solution and report the solubility.   Use a weighed sample of 0.10 g (weighed within ±0.01 g).   The known concentration is also of value in a gas-chromatographic examination.

Liquids. Use graduated pipettes to accurately measure the amount of liquids added to the test solvent.   If two colorless solutions are present, it may be difficult to see the boundary between the two layers.   Shake vigorously.   If two layers are present, the solution will become opalescent and become clear again as the droplets coalesce into two layers.   If both phases have similar refractive indices, inject a portion from the upper layer.   If both are identical, the solution is homogeneous. If different, the solutions are nonhomogeneous (see Fig. 3-5).

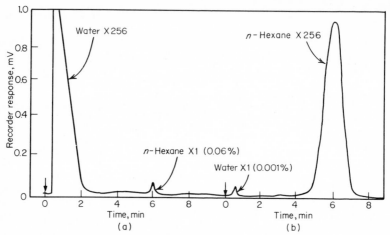

Fig. 3-5   Solubility of (*a*) *n*-hexane in water and (*b*) water in *n*-hexane determined on a Poropak Q column (3 ft × ¼ in.) at 125°C.

Do not heat unknown samples being examined for solubility in acid or base as samples may hydrolyze under the influence of heat.   Simply shake the sample vigorously with the acid or base and observe solubility within a few minutes.   Neutralization of the acid or base with careful cooling and injection into the gas chromatograph can verify the degree of solution.

If sample size is somewhat limited, several tests may be made on the same sample.   For example, if it is insoluble in water, using the same sample, add 1 ml of sodium hydroxide solution (20%) to the 3 ml of water.   Observe and record the solubility.   If the powdered sample can be well distributed in the water, divide and use half for sodium hydroxide solubility and half for hydrochloric acid solubility.   When the sample is very insoluble, remove the test solvents by filtration on filter paper, wash with water, and test with other solvents.   The solvent can be injected into the instrument before and after contacting the unknown.   If a peak does not appear, the unknown may be nonvolatile until a proper volatile derivative has been prepared.

When the solubility test is to be made in cold concentrated sulfuric or phosphoric acid, pour the acid into the test tube and add the unknown to the solvent.   Note whether any reaction takes place, as evidenced by the production of heat, gas, an insoluble precipitate, or a change in color.   If a gas is formed, collect it in a tube closed with a serum cap and inject a sample of the gas.   The precipitate, the solubilized derivative, or color derivative may be injected after suitable treatment to volatilize the compound.   For example, if it can be solubilized by sulfuric acid, e.g., sulfonation, the sulfonic acid may be separated and derivatized to volatile ester.   Repeat results, as suggested by Shriner et al.[40]

SOLUBILITY IN GAS CHROMATOGRAPHY

The solubility or nonsolubility of a substance in the liquid phase surrounding each particle in a gas-liquid chromatographic column is probably the most important physicochemical phenomenon of this technique.

The substance being analyzed is vaporized into the inert-gas stream. The sample is then carried into the column, and each component diffuses into the nonvolatile liquid phase coated on the column particles. The differences in solubility, distribution, volatility, and such other physiochemical phenomena as hydrogen bonding, boiling points, molecular structure, activities, etc., allow each component to separate to a greater or lesser extent by the time the sample reaches the end of the

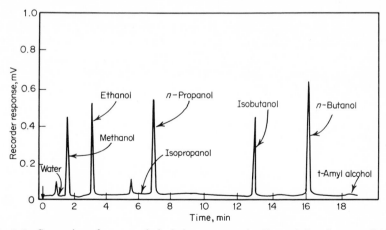

Fig. 3-6 Separation of water and alcohols on a nonpolar Poropak Q column at 170°C (6 ft × 3/16 in.).

column.   Since there are many texts on the theory, no attempt will be
made to present it here.   Reference is made to these books.[8,30,44]

To illustrate the differences in solubility in the liquid phase, water
may be separated before the alcohols on a nonpolar column (Fig. 3-6)
or after the alcohols on a polar column (Fig. 3-7).   A highly polar
column causes considerable tailing of the water as it is eluted (Fig. 3-8).

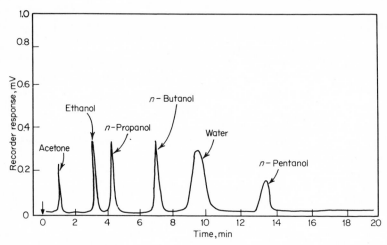

Fig. 3-7   Separation of water after *n*-butanol on a polar polyethylene glycol 300 column
at 100°C.[5]

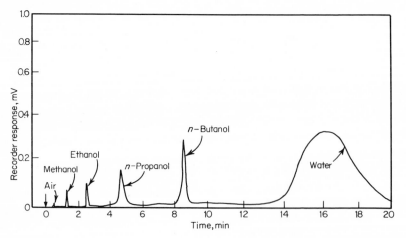

Fig. 3-8   Behavior of alcohols and water on a highly polar diglycerol column (10%)
(3 ft × ¼ in.) at 100°C.

GENERAL THEORETICAL CONSIDERATIONS

In the study of the solubility of compounds, it is possible through certain general rules to predict the probable solubility by looking at the structural formulas.   Also by observing the effects on various columns, it is possible to predict probable structural formulas.   As with all general rules, there are always exceptions.   However, in gas chromatography, the solubilities can be predicted with a much greater degree of precision as long as the solubilities are predicted from suitable homologous series (see Fig. 3-9).

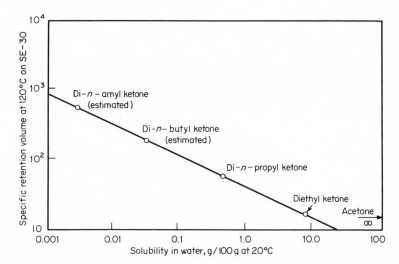

Fig. 3-9 Solubility of ketones in water at 20°C vs. specific retention volume on SE-30 (some values of solubility predicted).[25]

Solubility Related to Polarity

One well-known rule of solubility is that like dissolves like.   This applies to like-polarity solvents which dissolve like-polar compounds. In the solubility phenomenon, when an organic solid or liquid solute dissolves in a solvent, its molecules become randomly dispersed among the solvent molecules.   The dissolution of a substance that ionizes is somewhat different.   For example, in crystalline salt, sodium chloride, the average distance between the sodium ions and chloride ions is 2.8 Å. The average distance between ions in a molar solution is 10 Å.   When a substance is melted, the ions are moved farther apart by the action of heat.   Because of the highly polar character of sodium chloride, it must be heated to 800°C before the molecules melt and move further apart. Heating to 1413°C is required before sodium chloride vaporizes.

Considerable energy is necessary to move the ions of sodium chloride farther apart.

If sodium chloride is mixed with a solvent of low dielectric constant, such as decane (dielectric constant 1.99), or with a slightly higher polar compound, such as butanol (dielectric constant 17.1), little or no solution results. If the dielectric constant of the solvent is increased, the material dissolves more readily. It has been found that high-dielectric-constant solvents are necessary to effect solution. However, a solvent with too high a dielectric constant is ineffective. The solution of ionizing substances is quite complex. The most important principles for the good solvating powers of water and similar hydroxylated solvents is their ability to form bonds to oxygen via the hydrogens.[12]

A highly polar solvent with high dielectric constant like water does not dissolve a nonpolar substance because hydrogen bonds are not formed. As shown in Fig. 3-10, the water molecules arrange themselves in such

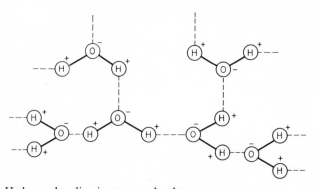

Fig. 3-10   Hydrogen bonding in water molecules.

a fashion that the induced positive and negative charges are adjacent to each other through hydrogen bonds. If an attempt is made to dissolve a nonionizing (nonpolar) substance such as decane (low dielectric constant) in a polar (ionizing) solvent such as water (high dielectric constant), no appreciable solution effect is found. Thus, a solvent of low dielectric strength (nonpolar), in most instances, will dissolve a substance of low dielectric constant (nonpolar). A polar solvent (high dielectric constant) similarly dissolves only polar solutes with high dielectric constants. "Like dissolves like" has been a good generalization.

Most organic molecules have a polar portion and a nonpolar portion. The solubility of molecules rests upon the strength (or length) of either of these portions. For example, the highly polar methanol or ethanol molecules are soluble in the polar solvent water. As the hydrocarbon portion of the molecule is increased, it takes on more of the properties of the hydrocarbon. Thus, *n*-octanol decreases in dielectric constant

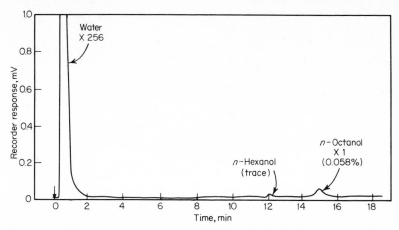

Fig. 3-11 Solubility of *n*-octanol in water at 25°C as determined on a Poropak Q column at 175°C (3 ft × ¼ in.).

and increases in ether solubility (Fig. 3-11). Similarly, if the aromatic hydrocarbon groups are increased on the polar portions of the molecule, the solubility in a polar solvent decreases. In the series phenol, 1-hydroxy-naphthalene, and *p*-phenylphenol, the solubility decreases in water and increases in ether. The aromatic radical, phenyl, is equivalent in solubility properties to about four aliphatic carbons when it is substituted in polar molecules such as aldehydes, alcohols, acids, and similar compounds. However, the retention data on polar columns in gas chromatography are different, depending not only on the solubility but also on volatility, boiling point, and other properties. For example, on a Carbowax 20M column, the retention data for amyl alcohol are considerably different from those for benzyl alcohol (see Fig. 3-12). Yet

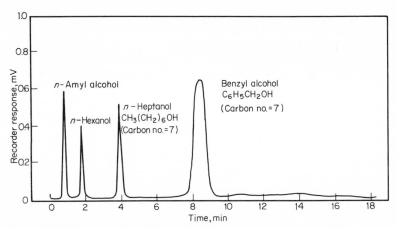

Fig. 3-12 Comparison of *n*-amyl alcohol and other homologs with benzyl alcohol on a polar Carbowax 20M column (3 ft × ¼ in., 10% at 150°C).

the solubility in the polar solvent water is almost the same.   Similarly, the water solubility of *n*-heptanoic acid is about equivalent to that of hydroxycinnamic acid, but the behavior on a polar gas-chromatographic column is quite different (see Fig. 3-13).

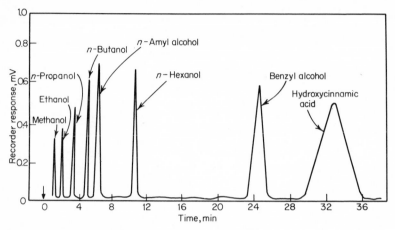

Fig. 3-13 Chromatogram of *n*-alkyl alcohols, benzyl alcohol, and hydroxycinnamic acid on 12 ft × ¼ in. Carbowax 20M column (20%) at 225°C.

Intermolecular Forces

The phenomenon of solubility is an equilibrium between the solid substance (or liquid or gas) and its solution.   Such an equilibrium is affected by interactions between the solvent and solute as well as the intermolecular forces of the solute.   It can be shown that these intermolecular forces do not depend upon solvent polarity or other solvent properties.   For example, the relative strengths of these forces may be partially estimated by comparing the melting and boiling points. Separating molecules by the process of melting a solid or boiling a liquid is somewhat similar to separating molecules when the substance goes into solution.

In Figs. 3-2 and 3-14, monocarboxylic acids show a slight variation of melting point and solubility.[6,25]   The dicarboxylic acids (Fig. 3-15) show, to a greater extent, the inverse relationship between melting point and solubility.   Each member with an even number of carbons melts at a higher temperature than the preceding or following dicarboxylic acid with an odd number of carbons.   The forces holding the crystals together (intercrystallinity) seem to be greater in molecules with an even number of carbons than in those with an odd number. This is probably due to the greater symmetry of the even-numbered carbon molecules.   Pimelic acid (with seven carbons) is more than twice as soluble as adipic acid (with six carbons). Even azelaic acid (nine carbons) is almost twice as soluble as suberic acid (eight carbons). As the center chain of the dicarboxylic acid lengthens, it approaches the properties of the hydrocarbons.

The dual properties of low solubility and high melting point occur in the isomers of maleic and fumaric acids also; see Table 3-1.

Generally, among cis-trans isomers, the trans form is usually less soluble and the cis form is more soluble in water. The reverse is true of the melting points: cis is lower and trans higher. The solubility in

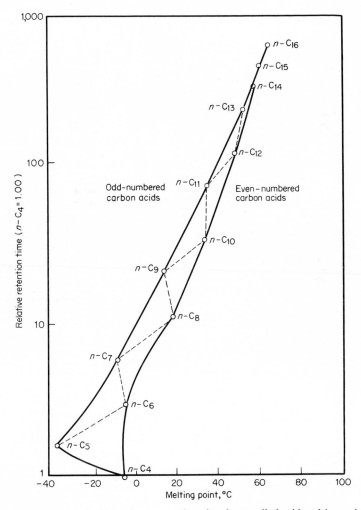

Fig. 3-14 Variation of odd- and even-numbered carbon n-alkyl acid melting points vs. their relative retention time (on silicone–15% stearic acid column) at 137°C.[6]

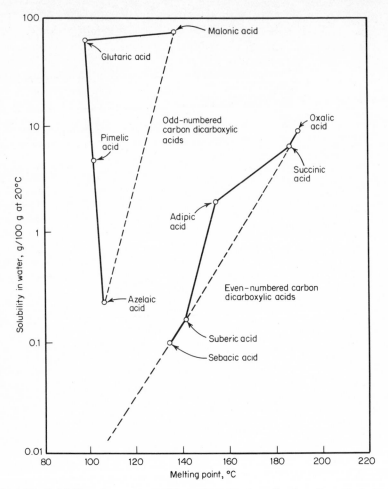

Fig. 3-15 Solubility of even- and odd-numbered carbon dicarboxylic acids in water vs. melting points.[39]

TABLE 3-1 Comparison of Isomers

	HC—COOH ‖ HC—COOH	HOOC—CH ‖ HC—COOH
	Maleic acid (cis)	Fumaric acid (trans)
Melting point	130°C	Sublimes 286°C
Solubility in water	Soluble (more than 3 g/100 g at 25°C)	Insoluble (0.63 g/100 g at 25°C)

water may be somewhat related to the ability to form hydrogen bonds with the water. Since both —COOH groups are on one side in the cis form, these bonds may form more readily than when they are on opposite sides, as in the trans form. Analogously, in such polymorphous compounds as benzophenone, the more soluble form has the lower melting point. Phthalic acid

melts at 206°C and is slightly soluble in water; terephthalic acid

melts at 300°C (sublimes) and is insoluble in water. Though these are not cis-trans isomers, the analogy is valid.

As a further illustration of the crystalline forces, of which the melting point is a good measure, consider the diamides of the dicarboxylic acids. Urea melts at 132°C and is water-soluble (it may be considered a diamide of formic acid). Oxamide has a high melting point, 420°C, and is only slightly soluble in water (see Table 3-2)).

TABLE 3-2

	NH_2 $\|$ $C{=}O$ $\|$ NH_2 Urea	NH_2 $\|$ $C{=}O$ $\|$ $C{=}O$ $\|$ NH_2 Oxamide	$NHCH_3$ $\|$ $C{=}O$ $\|$ $C{=}O$ $\|$ $NHCH_3$ *N,N′*-Dimethyl oxamide	$N(CH_3)_2$ $\|$ $C{=}O$ $\|$ $C{=}O$ $\|$ $N(CH_3)_2$ *N,N,N′,N′,*-Tetramethyl oxamide
M.p., °C	132	420	217	80
Solubility	Soluble (100 g/100 g)	Insoluble (less than 3 g/100 g)	Soluble (greater than 3 g/100 g)	Soluble (much greater than 3g/100 g)

Substituting methyl groups for the hydrogens in oxamide lowers the melting point and increases the solubility in water. The diamide of adipic acid (six carbons) is insoluble in water. When the tetraethyl derivative is made, it becomes water-soluble. Most stable amides can be observed without degradation on the gas chromatograph. Since the extremely high melting amides are difficult to volatilize in the gas-chromatographic columns, more volatile derivatives must be formed.

Amides of the general class $RCONH_2$ and $RCONHR'$ show a borderline solubility about a chain length of five carbons. The N,N-dialkyl amides ($RCONR'_2$) have lower melting points than the respective unsubstituted amides, resulting in a greater solubility in water. The limit of solubility between classes is around nine or ten carbons. Amides with the $-CONH_2$ group can show intermolecular hydrogen bonding (Fig. 3-16). Since this hydrogen bonding does not take place as readily in the N,N'-disubstituted amides ($RCONR'_2$), the molecules are less closely associated in the crystal, as demonstrated by lower melting points and a greater degree of solubility.

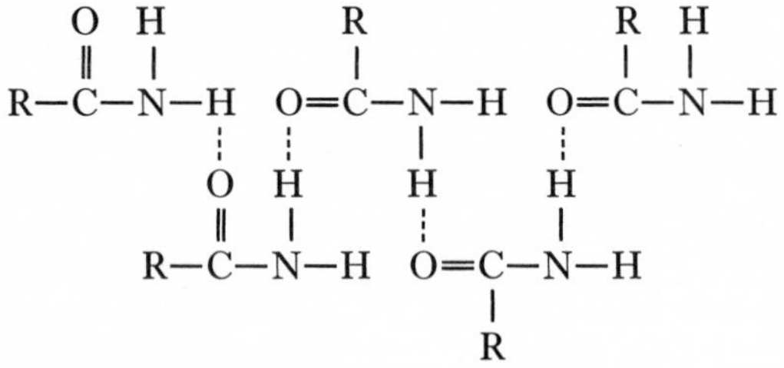

$$
\begin{array}{cccc}
\text{O} \;\; \text{H} & \text{R} & \text{R} \;\; \text{H} \\
\| \;\;\; | & | & | \;\;\; | \\
\text{R—C—N—H} & \text{O=C—N—H} & \text{O=C—N—H} \\
\vdots \;\;\; \vdots & | & \vdots \\
\text{O} \;\; \text{H} & \text{H} & \text{H} \\
\| \;\;\; | & \vdots & | \\
\text{R—C—N—H} & \text{O=C—N—H} \\
& | \\
& \text{R}
\end{array}
$$

Fig. 3-16 Hydrogen bonding in an amide.

If the molecular weight of a compound is increased, the intermolecular forces in the compound are correspondingly increased in the solid state. Polymers and other molecules of high molecular weight have lower solubilities in water and ether. For example, vinyl acetate monomer and monomeric formaldehyde are soluble in water; polymeric vinyl acetate and paraformaldehyde are insoluble. As shown in Fig. 3-17, poly(methyl acrylate) depolymerizes to re-form certain amounts of monomer.

$$x \; CH_2O \longrightarrow HO(CH_2O-)_xH$$

(Water-soluble)   (Water-insoluble)
Formaldehyde   Paraformaldehyde

The polymers in each instance are insoluble in water, but the monomers are water soluble:

$$x \; CH_2{=}CHCOOCH_3 \xrightarrow{\text{polymerize}} -\left[\begin{array}{c} CH_2{-}CH \\ | \\ COOCH_3 \end{array}\right]_x \xrightarrow{\text{depolymerize}}$$

Soluble   Insoluble

$$X \; CH_2{=}CHCOOCH_3$$

Soluble

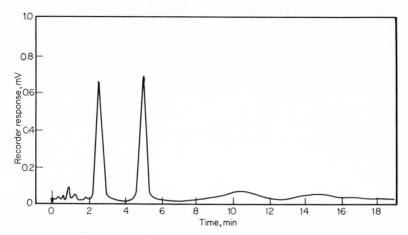

Fig. 3-17 Separation of monomers by pyrolysis (at 450°C) of poly(methyl acrylate) analyzed on a diethylene glycol succinate column (24%, 12 ft × ¼ in.) at 140°C.[11,27]

There are many examples of monomers that are water-soluble and polymers that are insoluble, especially with biochemical materials. Glucose, for example, is water-soluble, but its polymers (starch, cellulose, glycogen, etc.) are insoluble. Most of the amino acids are soluble, but their polymers, the proteins, are insoluble. Certain biochemical polymers, such as proteins, dextrins, starches, etc., tend to form colloidal suspensions that superficially appear to be water-soluble.

Introduction of a halogen into the molecule increases the molecular weight and reduces water solubility. If the number of halogens in a given molecule is increased, its ether solubility is reduced, but usually not enough to make it insoluble in ether. Retention time in gas chromatography lengthens as the molecular weight is increased by halogen substitution or by more substitution.[10]

Solubility Classes Applied to
Gas Chromatography

In studying an unknown substance, important information can be obtained from its solubility in various solvents.

1. A functional group may be found. For example, butyl alcohol is partially soluble in water; butane is almost insoluble in water. The partial solubility of the butyl alcohol in the water indicates that a functional group is present. Even slight solubilities can be measured by gas chromatography and serve as clues to the presence of functional groups (see Fig. 3-18).

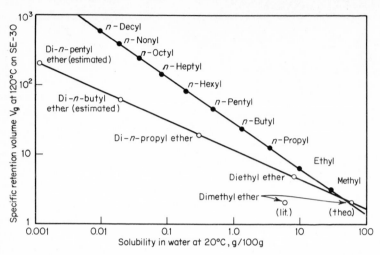

Fig. 3-18 Solubility of ethers and formate esters in water at 20°C related to their specific retention volumes on SE-30 at 120°C.[23]

The procedures of Kamm[20] and Shriner, Fuson, and Curtin,[38] which are similar, are summarized in Fig. 3-3. They have been given in greater detail by Cheronis and Entrikin[7] and further expanded to include examination of each layer by gas chromatography. Thus, the actual solubility of aniline in water can be determined as well as the solubility in other specific solvents. Also indications of molecular size can be obtained from these data. The solubilities give leads to the identity and the gas-chromatographic information assists in confirming these findings.

2. Certain solubilities in specific solvents give more definite information about the functional groups. For example, aniline is practically insoluble in water and sodium hydroxide solution (5%) but soluble in hydrochloric acid solution (5%), which converts it into a soluble salt, aniline hydrochloride. This action is a good indication of a basic functional group.

3. Certain conclusions about molecular size can be obtained from a solubility study. Generally, for this study, solubilities under 3 percent are considered insoluble and above 3 percent soluble. In many homologous series with single functional groups, compounds with more than five carbon atoms are usually insoluble; those with less than five carbons are soluble in water. However, with gas chromatography, the exact solubility can be determined, as shown in Fig. 3-18. From the plot of a given homologous series, the molecular weight can be predicted quite accurately.

Effect of Chain Branching

Chain branching of the lower homologous series of hydrocarbons, alcohols, etc., reduces the intermolecular forces and decreases the inter-molecular attractions. As a result, the solubility is increased and the melting points are decreased compared with the corresponding straight-chain compounds. Usually, the retention times are also reduced in relation to the corresponding straight-chain substances (see Fig. 3-19). For example, the branched compounds of the same number of carbon molecules usually appear in the chromatogram before the molecules of the straight-chain compound. The solubility of an iso compound (like isobutanol) is much greater than its normal isomer and is closer to the solubility of the lower normal homolog. Shriner et al.[42] have shown the effect of chain branching on the solubility of various classes of compounds. The most highly branched of a series of isomeric com-pounds generally has the greater solubility, lower melting point, and lower chromatographic retention time.

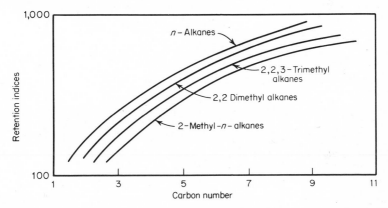

Fig. 3-19 Carbon numbers of *n*-alkanes and branched alkanes vs. retention indices (on SF-96, silicone oil) at 25°C.[15]

It can be demonstrated also that the position of the functional group along the carbon chain has an effect not only on the solubility but also on the retention time and melting and boiling points. As an illustra-tion, for a member of a homologous series, 1-pentanol is least soluble in water and has the highest boiling point and the longest retention values; 2-pentanol is more soluble and has a lower boiling point and retention value; and 3-pentanol is the most soluble, having the lowest boiling point and lowest retention value. Combining the effects of branching and of moving the functional group toward the center of the molecule gives a marked increase in solubility. The boiling points are decreased and the retention values reduced, as noted for 2-methyl-2-butanol.   The more symmetrical or compact the structure, the greater its solubility (providing comparisons are made on the same types).

Effect of Molecular Structure on Acidity or Basicity

Usually a water-insoluble compound dissolves in dilute acid or base, depending upon its basic or acidic strength.   The structural details that determine whether a substance is acidic or basic depend upon several principles.   These factors also affect the behavior on chromatographic columns and other physiochemical properties.

Electronic Effects. Jaffe[19] reported on some correlations between

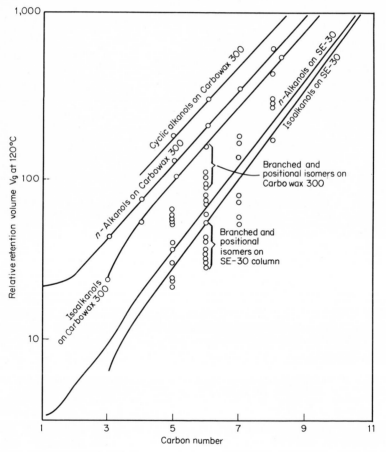

Fig. 3-20   Behavior of alkanols on polar Carbowax 300 and nonpolar silicone gum rubber, SE-30, at 120°C.[26,28,46]

quantitative structural relations and the acidic or basic strength of substituted aromatic compounds. These effects on the acidity or basicity have been shown to be mostly electronic. Certain groups have properties of withdrawing electrons or increasing electrons around the benzene ring.[31] Taft[45] made quantitative correlations between dissociation constants in aliphatic acids and polar and steric effects. Certain chromatographic columns have a greater effect on a compound according as it is acidic or basic (see Fig. 3-20).

Almost all organic carboxylic acids are fairly strong acids (dissociation constants at 25°C in water are about 10^{-6} or less). These are easily soluble in sodium bicarbonate (5%) and sodium hydroxide (5%) solutions. Phenols (but not carboxylic acids) have acidic hydrogens due to the benzene-ring withdrawal of electrons (dissociation constant about 10^{-10}), but they are much weaker acids and are not soluble in sodium bicarbonate solution (5%) (first dissociation constant 4×10^{-7}), though they are soluble in sodium hydroxide (5%) solution. If a substituent group is introduced onto the benzene ring, the effect on the acidity of the phenol is dramatic. For example, when a nitro group ortho or para to the phenol is substituted, the dissociation constant increases 600 times, to about 6×10^{-8}. The compound is still soluble in sodium hydroxide solution (5%) but not in sodium bicarbonate solution (5%). If two nitro groups are introduced, as in 2,4-dinitrophenol, the compound becomes soluble in both reagents due to its increased acidity. A similar effect can be demonstrated in the reduction of aniline basicity by nitration. The ability of the nitro group to increase acidity is due to its electron-withdrawing effect, which stabilizes the phenoxide anion. The negative charge on the nitro group is partially distributed:

An identical increase in the acidic properties of phenol is observed when halogens are introduced. If a bromine is added ortho, the acidity is increased about thirtyfold. A para bromine increases its acidity by a factor of 5. A 2,4,6-tribromophenol is acidic enough to be soluble in sodium bicarbonate solution (5%) and sodium hydroxide solution (5%).

Bromine evidently stabilizes the tribromophenoxide anion, possibly due to the inductive effect of this element. There is a greater probability for resonance stabilization in the nitrophenoxide anions than in the

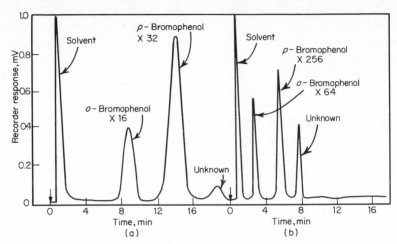

Fig. 3-21 Behavior of bromophenols on (*a*) a polar column (Carbowax 20M, 5%, 3 ft × ¼ in. at 200°C) and (*b*) a nonpolar column (Apiezon L, 10%, 3% KOH, 3 ft × ¼ in. at 150°C).

bromophenoxide anions.[17]   The bromophenol behaves differently on columns of differing degrees of acidity (see Fig. 3-21).   When the tri-methylsilyl derivative is prepared, this effect is masked, and the ions behave in accordance with their molecular weight.

Basicity of amines can be affected by similar electronic influences. Aliphatic amines have hydrolysis constants of about 10^{-3} or 10^{-4} (ammonia is 10^{-5}).   Substituting the benzene ring for the aliphatic chain lowers the K_b to 5×10^{-10} (aniline).   The phenyl group stabilizes the amine on the left of equilibrium by resonance:

$$H_2O + \left[\begin{array}{c} H-N-H \\ \bigcirc \end{array} \longleftrightarrow \begin{array}{c} H-N-H \\ \bigcirc \\ H^- \end{array} \right] \rightleftharpoons \begin{array}{c} H^+ \\ | \\ H-N-H \\ \bigcirc \end{array} + OH^-$$

Introduction of a second phenyl group decreases the basicity so much that the compound has no measurable alkalinity in water.   Diphenylamine is so nonbasic that it is insoluble even in dilute hydrochloric acid solution (5%).   If a nitro group is substituted on the benzene ring of aniline, the basicity is reduced by the electron-withdrawing effect of this group.   The structure responsible is probably the resonance hybrid:

H—N—H

(quinoid ring structure)

$^-$:Ö—N—Ö:$^-$

As the basicity or polarity is reduced, less retention or holdup on polar columns results. There is no appreciable effect on nonpolar columns.

Steric Effects. There is a noticeable reduction in solubility of ortho-substituted phenols in sodium hydroxide solution (5%). It is necessary to increase the alkali to 35% and use potassium hydroxide in methanol-water mixture to effect solution of these sterically hindered phenols. If groups on either side of the phenol group are substituted, the solubility is reduced further. The more branched the substituent groups, the greater the steric hindrance and the greater the reduction in solubility. For example, 2,4,6-tri(*t*-butyl)phenol is. insoluble in the 35% alkali solution.[43] It can be made into a sodium salt only by reaction with elemental sodium in liquid ammonia.

Steric hindrance can be demonstrated in many compounds. For instance, 2,4,6-tri(*t*-butyl)aniline has such weak basicity that it is nearly impossible to measure in water.[1] Similarly, 2,6-di(*t*-butyl)-pyridine is a weaker base than 2,4-dimethylpyridine.[4] Several theories have been postulated for the reduction in basicity of the amines, the most probable being that of steric or spatial interference with the hydrogen bonding to solvate the ion.[16] As shown in Fig. 3-22, the branched substituents not ortho to the functional group are retained to a greater extent in a chromatograph than the corresponding ortho substituents. The effect is most pronounced in the disubstituted compounds when both are ortho to the functional group versus those not ortho.

On carboxylic acids, steric effects can either increase or decrease the acidity. In aromatic acids, the ortho group increases the acidity. Ortho-substituted benzoic acids are stronger acids than the para isomers, probably because steric forcing of the carboxylic group out of the plane of the benzene ring allows better solvation of the group[18] and produces a kind of steric inhibition of resonance.[47] If nitro groups are substituted in phenol para to the hydroxyl group, the pK_a at 25°C changes from 9.99 to 7.21 (in *p*-nitrophenol), increasing the acidity by 2.8 pK_a units. Adding two methyl groups meta to the hydroxyl increases the pK_a only to 8.24, or 1.7 pK_a units stronger than phenol. This effect is due in part to the electronic reduction produced by the methyl groups and to a

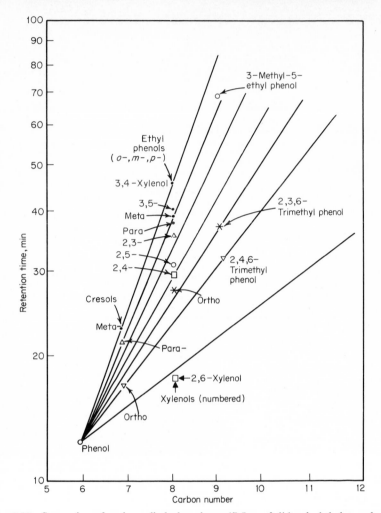

Fig. 3-22 Separation of various alkyl phenols on 47.5 m of didecyl phthalate column at 120°C.[21]

greater extent to the steric inhibition of resonance (Fig. 3-23). For structure *a* to contribute significantly to the acidity due to the electron-withdrawing ability of the nitro group, the nitrogen and the benzene ring must be in or near the same plane. The introduction of methyl groups in *b* inhibits this planar arrangement, and hence the acidity is less.

In aliphatic carboxylic acids, if alkyl groups are substituted on the carbon α to the functional group, the tendency is to decrease acidity, due probably to steric hindrance of hydrogen bonding (or solvation) of

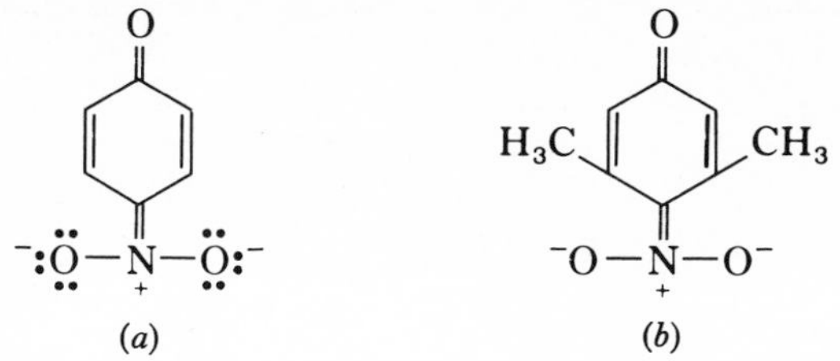

Fig. 3-23 Effect of branching on acidity: (*a*) greater and (*b*) less acidity.

the carboxylate group with the water.[14] Substitution of electron-withdrawing groups (as $-NO_2$ and halogens) has the reverse effect by increasing the acidity. The steric effect is greatly reduced or eliminated by moving the substituent alkyl group farther from the carboxylic acid.

Since carboxylic acids have strongly polar groups, they are retained more readily on polar chromatographic columns. Chromatographing free acids on columns with free hydroxyl groups is not recommended because of their tendency to esterify with the column coating. Similarly, ester exchange may occur if free acids are chromatographed on polyester columns. Amines usually can be chromatographed on all but strongly acidic columns.

SOLUBILITY REAGENTS

Water

In most solubility studies, like dissolves like. Hence, highly polar water dissolves polar substances. Hydrocarbons, which are nonpolar, would not be expected to have appreciable water solubility. If unsaturation, aromatic groups, or halogens are introduced into the hydrocarbons, the polarity is not greatly altered and the solubility in water is not greatly different from that of the parent compound. However, there is a slight tendency to hydrogen-bond with chlorine. Introduction of halogens increases molecular weight, and water solubility decreases over the parent compound with the increasing molecular weight.

Salts are extremely polar and generally water-soluble. There are exceptions to this rule. Inorganic salts of organic compounds usually cannot be chromatographed without conversion to volatile esters. For example, the sodium salts of C_1 to C_6 *n*-alkyl acids are nonvolatile. When converted to the methyl or ethyl esters, they become volatile and are easily chromatographed (see Fig. 3-24).

Since alcohols, esters, ethers, acids, amines, nitriles, amides, ketones, and aldehydes have varying degrees of polarity, they would be expected to have appreciable water solubility. As shown in Fig. 3-2, the lower

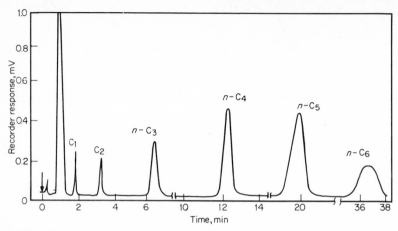

Fig. 3-24 Separation of methyl esters of n-alkyl carboxylic acids on 6 ft $\times$ $3/16$ in. SE-30-XE-60 (2.5/0.5%) on Anachrom AS, 80/90 mesh at 75°C.[11]

alcohols, being highly polar, have considerable solubility in water. As the hydrocarbon chain begins to dominate the molecule and dilutes the polar effect, the solubility decreases. Pentanol is soluble to the extent of 2 g/100 g of water, while hexanol is soluble only to the extent of 0.6 g/100 g of water.

Acids and amines are usually more water-soluble than the correspond-ing neutral compounds, which tend to form complex hydrogen-bonded substances with the water. This observation is further substantiated by the fact that the solubility of amines decreases as the basicity decreas-es. In acids the solubility is reduced as the acidity decreases. Tertiary amines, which are soluble in cold water but not in hot water, form stable hydrates that are soluble at the lower temperatures and are unstable at the higher temperatures. The solubility illustrated here is that of the free amine rather than the hydrate. Acids similarly owe their solubility to the hydrogen-bonding tendency and hydrolysis of the hydrogen on the carboxyl group.

In the monofunctional acids, alcohols, aldehydes, amides, esters, ethers, ketones, and nitriles, the lower limit of water solubility in a given homologous series may be found around the C_5 molecule. For con-venience, the lower limit has been established as 3 percent soluble, anything less soluble than 3 percent being considered insoluble. However, the actual determination of the degree of solubility by gas chromatography (even above or below this arbitrary limit) can assist the investigator in determining the presence of a functional group and pos-sibly the length of the carbon chain, etc. (see Fig. 3-1).

It has been shown that as the chemical and structural similarity between the solute and solvent is increased, the mutual solubilities are increased. Since water is extremely polar, solubility in water requires the presence of polar groups. In any given homologous series containing a polar group, as the hydrocarbon portion of the molecules (nonpolar) is increased, the polar portion remains unchanged, but a steady decrease in solubility in the polar solvent nevertheless results from dilution of the polar portion of the molecule by the nonpolar hydrocarbon chain.

The polarities of many functional groups are quite similar. The dilution effect of the hydrocarbon chain is similar for most series, in that approximately the same length of chain produces the same reduction in solubilities. As noted, for the arbitrary limit of 3 percent it takes approximately five carbons in the chain. If an arbitrary limit of 1 percent, 0.3 percent, or some other level were chosen, probably the carbon chain length would be different, but it would still be similar.

Determining solubility by gas chromatography frees the investigator from any arbitrary limit except for the convenience of the solubility scheme presented by Shriner, Fuson, and Curtin.[38] Thus, in studying the solubility of ethers in water (Fig. 3-19) di(n-propyl) ether is soluble to the extent of approximately 0.55 g/100 g water (estimated); di(n-butyl) ether is soluble 0.05 g/100 g, whereas diisobutyl ether is less soluble due to steric hindrance of hydrogen bonding. The length of chain or degree of branching or both can be established by a study of the solubility of ethers in water.

Certain oxygenated compounds tend to form complexes with water, such as hydrates, thus increasing their solubility above their predicted values. If these hydrates are quite stable, the compounds tend to be very soluble in water. Chloral hydrate is an example. Such compounds also tend to dehydrate in the injection port of a chromatograph but may be strongly retarded on polar columns due to hydrogen bonding. Figure 3-25 illustrates the separation of chloral hydrate on a polar and a nonpolar column.

Water solutions of such low-molecular-weight esters as ethyl formate and methyl pyruvate have a tendency to turn acidic due to hydrolysis at room temperature. Most of these esters are soluble in ether without any discernible reaction. Shriner et al.[39] have compared the water solubilities of even- and odd-numbered dicarboxylic acids.

Ether

Ether and water are considered polar solvents, based on the classification of dipole moments. Ether, however, differs greatly from water in its effect on polar compounds. Since ether alone does not hydrogen-bond or associate with itself as water does, it has a low dielectric constant (4.3).

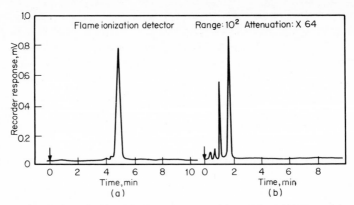

Fig. 3-25 Separation of chloral hydrate (commercial grade) at 100°C on (a) a polar column (Carbowax 20M, 10%, 3 ft × ¼ in.) and (b) a nonpolar column (Apiezon L 10%, 3 ft × ¼ in.).

Water has a very high value (80) for such a low-molecular-weight compound. Nonpolar compounds dissolve in ether but do not dissolve in water. Ionic substances, such as salts of acids and amines, do not dissolve in ether but dissolve readily in water. Ether can form hydrogen bonds only through its oxygen. Water can form hydrogen bonds through either its oxygen or its hydrogen.

The solubility in ether of water-soluble compounds depends not only on the number of functional groups but also on their character. Polarity of a compound is due to an unsymmetrical charge distribution along the molecule. For example, since hydrocarbons have a uniform charge distribution, they are nonpolar and soluble in ether. If one hydrogen is replaced with a polar group, unsymmetrical charges along the molecule are produced, inducing a dipole and reducing the ether solubility. The water solubility may be increased. If more than one functional polar group is present, ether solubility is decreased even further and water solubility is increased.

In polar compounds such as acids, the hydrocarbon portion of the molecule is soluble in ether but insoluble in water. If the hydrocarbon portion is not too dominant over the polar portion of the molecule, the compound will be soluble in both ether and water. When the hydrocarbon portion becomes dominant, the ether solubility increases and the water solubility decreases. (See Figs. 3-2 and 3-26 for the relationship of water and ether solubilities to retention times.) When butane and decane, which are nonpolar and soluble in ether but insoluble in water, are converted into the corresponding carboxylic acids, polar compounds are produced. The acid groups of both have about the same affinity for water and reduced solubility in ether, but the hydrocarbon portions are soluble in water to different extents and probably show different solubilities in ether at low temperatures.   Decanoic acid is soluble in ether (0.003 g/100 g water).   Butanoic acid is also soluble in ether, but probably much less at low temperatures, and in water (Fig. 3-26).

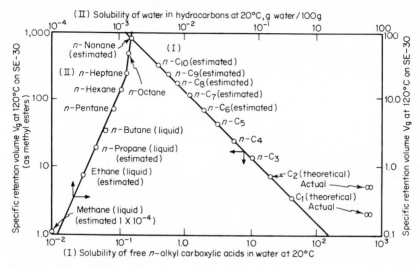

Fig. 3-26   The solubility of free fatty acids and hydrocarbons (*n*-alkyl) in water at 20°C vs. the specific retention volume of their corresponding methyl esters on SE-30 column at 120°C.[23]

Slightly ionized compounds with only one polar group are soluble in ether.   Sulfonic acids and their salts are highly polar and thus insoluble in ether.

Solubility of different isomers of the same compound in ether sometimes differs markedly and provides a useful method of separation.

Commercial ether usually contains small amounts of alcohol and water as preservatives.   These components usually increase the solubility of compounds above expected values, especially water- and alcohol-soluble compounds that are only slightly soluble in ether.   For more accurate solubility studies, the ether can be shaken with water to remove the alcohol and then dried with calcium sulfate or molecular sieves. The solvent extraction of a solute from water by the ether or from ether by water (actually a phase of ether saturated with water on top and a lower layer of water saturated with ether) results.   Chromatographic examination of both phases can give the proportions of each (see Fig. 3-27).   For example, in attempting to extract benzoic acid from a water solution by ether, much less acid is extracted than theoretically calculated from the law of distribution if the calculations are based on

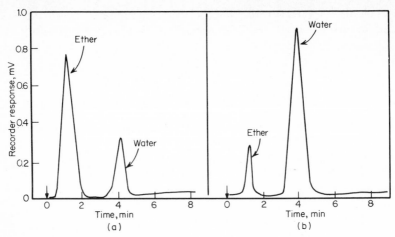

Fig. 3-27 Separation of ether and water in (*a*) ether layer and (*b*) water layer using a Poropak Q column (3 ft × ¼ in.) at 100°C.

pure water and pure ether.   The chromatogram (Fig. 3-28) shows that benzoic acid is more soluble in a saturated solution of ether in water than it is in pure water and less soluble in a saturated solution of water in ether than it is in pure ether.

If there is doubt about the solubility in ether, the ether solution can be injected into the instrument; evidence of solubility is checked by the appearance of a peak or of no solubility by the lack of a peak.   Verification of solubility or nonsolubility can be made in benzene.

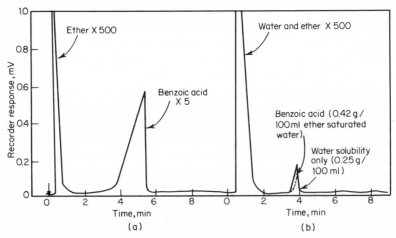

Fig. 3-28 Distribution of benzoic acid between (*a*) ether layer (66 g/100 ml solubility) and (*b*) water layer as determined on a W-98 (silicone gum rubber, 10%) 6 ft × ⅛ in. at 125°C. (*Note:* Reduction of the retention time is due to the effect of the water.)

Using gas-chromatographic data, as shown in Fig. 3-26, values for water solubility in hydrocarbons extrapolated well beyond the observable values may be applied to the solubility data of *n*-alkyl amines and di(*n*-alkyl) amines, as shown in Fig. 3-29. For example, Table 3-3

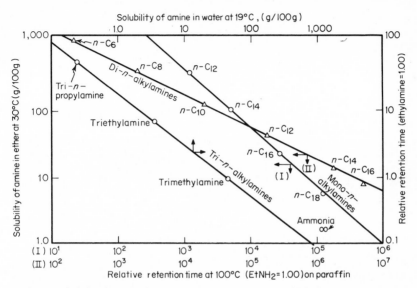

Fig. 3-29 (*a*) Solubility of tri(*n*-alkyl) amines in water at 19°C vs. relative retention time on paraffin oil column at 100°C.[2] (*b*) Relationship between the relative retention volume (on paraffin at 100°C) and the solubility of amines in water at 25°C.[2,33]

TABLE 3-3   Solubility of Amines in Ether at 30°C[29]

Amine	Solubility, g/100 g	Relative retention values
n-Decylamine	Infinitely soluble	350
n-Dodecylamine	275	1,500
n-Tetradecylamine	71	6,500
n-Hexadecylamine......	18.5	30,000
n-Octadecylamine	4.4	130,000*
Di(*n*-hexyl)amine	Very soluble	240
Di(*n*-octyl)amine	275	2,400
Di(*n*-decyl)amine	93	22,500
Di(*n*-dodecyl)amine	32.5	200,000
Di(*n*-tetradecyl)amine...	10.5	2,000,000*
Di(*n*-hexadecyl)amine...	6.8	5,000,000*
Di(*n*-octadecyl)amine ...	0.3	

* Numerical values are theoretical or extrapolated. (In actual practice high retention samples are rerun at higher temperatures.)

gives the literature values for the solubility of the amines in ether. Starred values were obtained from the plot of the hypothetical relative retention values versus solubility (see Fig. 3-30).

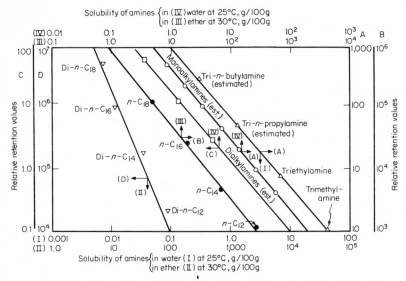

Fig. 3-30 Relationship of the solubility of *n*-alkyl and di- and tri- (*n*-alkyl) amines in water at 25°C and ether at 30°C to their relative retention values (on SE-30 at 120°C).[29]

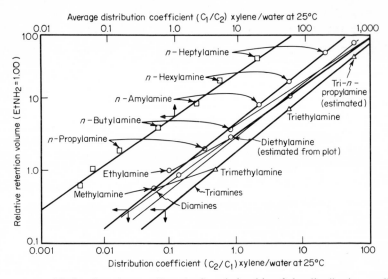

Fig. 3-31 *n*-Alkyl amines (mono, di, and tri): relationship of the distribution coefficient between xylene and water at 25°C to their relative retention volume (on paraffin oil at 100°C).[2,34]

Using a similar plot of the distribution coefficient of amines between water and xylene at 25°C (total of $C_2 + C_1 = 10$ mmol/l) versus the relative retention volume of the amines on paraffin at 100°C gives straight-line relationships (Fig. 3-31). The only values that do not fall on the homologous-series line are those for ethylamine and diethylamine, which *may* be slightly erroneous in their distribution coefficient. (Estimated corrected coefficient $C_2/C_1 = 0.100$ for ethylamine and 0.750 for diethylamine.)

Many physical properties can be checked by this technique. Values suspected of being inaccurate in any homologous series can be verified. Data beyond the measurable ranges can be extrapolated. Chapter 4 shows how the analyst can measure these physical properties even when the sample is too small to be measured directly; i.e., value can be predicted with a high degree of accuracy.

Dilute Hydrochloric Acid

For a complete discussion of hydrochloric acid solubility see the section on amide and amine solubility below.

Cold Concentrated Sulfuric Acid

This reagent is used on neutral, water-insoluble compounds as they do not belong to the miscellaneous class. Unsaturated compounds easily sulfonate; functional groups containing oxygen dissolve readily in cold concentrated sulfuric acid, which may be an indication of a reaction such as sulfonation, dehydration, polymerization, or addition. In addition reactions, the original substance can be recovered by dilution with ice water[13] (see Table 3-4).

Saturated aliphatics, paraffins, cycloparaffins, and their halogen compounds are insoluble in cold concentrated sulfuric acid and belong to the inert-compound class. Unsubstituted aromatic hydrocarbons and their halogen compounds are insoluble in this reagent, as they do not sulfonate at low temperatures. Insertion of two or more alkyl groups into the benzene ring makes the compound sulfonate more easily at room temperatures. The polyalkyl benzenes dissolve readily in cold concentrated sulfuric acid.

Sometimes the sulfuric acid reacts to produce an insoluble substance. For convenience, these substances are considered soluble in the sulfuric acid even though their reaction products are not soluble. This may assist in separations from mixtures.

Certain high-molecular-weight ethers sulfonate so slowly at room temperature that they are not considered soluble in this reagent. However, the neutralized acid may be esterified and injected into the

TABLE 3-4   Reactions of Organic Compounds with Sulfuric Acid*

Addition

$$RCH{=}CHR + H_2SO_4 \longrightarrow \left[RCH_2{-}\overset{+}{C}HR \right] \longrightarrow RCH_2\overset{\displaystyle OSO_3H}{\underset{\displaystyle |}{C}}HR$$

$$C_6H_5COOH + H_2SO_4 \longrightarrow C_6H_5\underset{\displaystyle \diagdown OH}{C}{-}OH + HSO_4^-$$

$$(C_6H_5)_2C{=}O + H_2SO_4 \longrightarrow (C_6H_5)_2COH + HSO_4^-$$

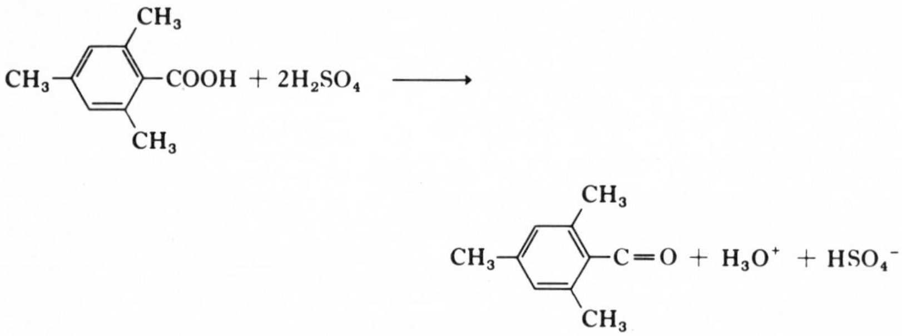

$$CH_3{-}\overset{CH_3}{\underset{CH_3}{\bigcirc}}{-}COOH + 2H_2SO_4 \longrightarrow$$

$$CH_3{-}\overset{CH_3}{\underset{CH_3}{\bigcirc}}{-}C{=}O + H_3O^+ + HSO_4^-$$

Dehydration

$$(RCH_2)_2COHCH_2R' + H_2SO_4 \longrightarrow (RCH_2)_2C{=}CHR' + H_3O^+ + HSO_4^-$$

$$2CH_3CH_2OH + H_2SO_4 \longrightarrow CH_3CH_2{-}O{-}CH_2CH_3 + H_3O^+ + HSO_4^-$$

Hydrolysis

$$CH_3\underset{\displaystyle O}{\overset{\displaystyle \|}{C}}{-}O{-}\underset{\displaystyle O}{\overset{\displaystyle \|}{C}}{-}CH_3 + 3H_2SO_4 \longrightarrow 2CH_3C(OH)_2 + HSO_4^- + HS_2O_7^-$$

Sulfation

$$RCH_2OH + 2H_2SO_4 \longrightarrow RCH_2OSO_3H + H_3O^+ + HSO_4^-$$

$$(C_6H_5)_3COH + 2H_2SO_4 \longrightarrow (C_6H_5)_3C^+ + 2HSO_4^- + H_3O^+$$

Sulfonation

$$CH_3O{-}\bigcirc + H_2SO_4 \longrightarrow CH_3O{-}\bigcirc{-}SO_3H + H_3O^+ + HSO_4^-$$

---

* R can be either the same or different groups.

chromatograph to assist in identification even though only a small amount is reacted.

Numerous secondary and tertiary alcohols dehydrate readily in cold concentrated sulfuric acid, yielding olefins, which may then polymerize. These polymers may be insoluble in the cold acid, forming a layer on its surface. For convenience, these alcohols are considered soluble in this reagent. Similarly, benzyl alcohol and its derivatives, which undergo condensation in the acid, forming colored insoluble products, are also classed as soluble in this reagent.

If any of the reaction products are to be examined by the instrument, it is necessary to neutralize the acid (to recover the product) or wash the product free of the acid. When the unknown is sulfonated, an ester must be prepared for chromatographing. Addition complexes can usually be recovered unchanged and chromatographed as is after washing.

Phosphoric Acid (85%)

Alcohols, aldehydes, cyclic ketones, methyl ketones, and esters readily dissolve in phosphoric acid (85%) if they contain less than nine carbon atoms. Ethers have a lower limit of solubility than nine carbon atoms in the molecule. Diethyl ether is soluble, but di(n-butyl) ether and anisole are not soluble. Certain olefins, such as amylene, are soluble in this reagent; ethyl benzoate is insoluble. Acetophenone, ethyl oxalate, and other compounds form solid substances with phosphoric acid (85%). This reagent dissolves organic compounds without producing appreciable quantities of heat or color (sulfuric acid produces both).

Substances that dissolve in phosphoric acid can often be recovered without appreciable change by dilution with water or neutralization of the acid. This technique is useful for separation of mixtures. The separated component may be injected into the instrument or if it reacts with the acid, esterified and chromatographed.

Shriner et al.[38] have tabulated various compounds to show borderline solubilities and classes.

SOLUBILITY OF AMIDES AND AMINES

Primary n-alkyl amines are soluble in water up to C_5 alkyl chain length (as shown in Fig. 3-31). In the secondary amines, the C_5 alkyl amine becomes less soluble. With tertiary amines, the tri(n-butyl)alkyl chain (C_4) is insoluble and the tri(n-propyl)amine (C_3) is only slightly soluble. Since these insoluble or slightly soluble amines form salts with acids, including hydrochloric acid, this reagent makes aliphatic amines more readily soluble.

If the phenyl groups are introduced into the amine, the basicity of the amine is reduced due to the electron-attracting ability of the aryl group. Even though phenylamine is weaker than the primary alkyl amines, it is still soluble in dilute hydrochloric acid.   Two aryl groups on the amine weaken the basicity so that it is no longer soluble in the dilute hydrochloric acid solution (5%).   The triaryl amine and carbazole are also less soluble in dilute acid.   The introduction of an aryl group into a secondary or tertiary amine does not reduce its solubility in hydrochloric acid unless more than one aryl group is introduced to reduce the basicity nearly to neutral.

Most amines chromatograph readily on columns treated with strong bases such as potassium hydroxide.   Figure 3-32 shows a comparison of retention times of amines on various columns.   Slightly acidic columns tend to retain the more basic amines strongly.

When the diamides, $RCONR_2$, are of high enough molecular weight to become water-insoluble, they are still soluble in dilute hydrochloric acid (5%).   The amides, $RCONH_2$, are essentially neutral compounds, as are the monosubstituted amides, $RCONHR'$, being insoluble in water and dilute hydrochloric acid (5%).   N-Benzylacetamide is an exception, as it is soluble in dilute hydrochloric acid (5%) and insoluble in water.

Occasionally, certain basic compounds will react with the dilute hydrochloric acid (5%) to form an insoluble salt.   Usually these are considered soluble even if their salts are insoluble or soluble only when warmed or diluted with water, such as the naphthylamines.   If in doubt whether a reaction has taken place, compare the melting points of the original and the reacted product or make a halogen test with alcoholic silver nitrate.   Inorganic salts of amines do not chromatograph readily without decomposition.   Occasionally organic salts of amines will chromatograph unchanged.   Others decompose in the instrument into amides that are volatile enough to be chromatographed.

Ammonia is considered the most basic of the amines.   As various groups are substituted for the hydrogens in the ammonia, these groups have degrees of electron-attracting power (electronegativity) as follows:

Least electron-attracting $\longrightarrow$ most electron-attracting

Alkyl $<$ aryl $<$ acyl $<$ aroyl $<$ sulfonyl

Increasing electronegativity $\longrightarrow$

The alkyl group has about the same electron-attracting power as one of the hydrogens.   Primary, secondary, and tertiary amines have about the same basicity as ammonia ($K_b = 10^{-3}$ to 10^{-5}).   Since the aryl group attracts more electrons from the nitrogen group, these amines are

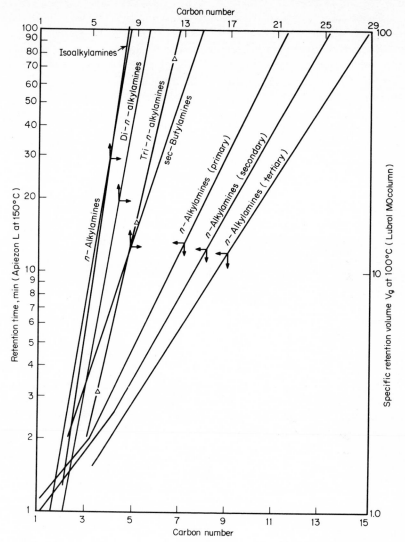

Fig. 3-32 Separation of amines (primary, secondary, and tertiary) on an Apiezon L column (5%, treated with 3% KOH) at 150°C (6 ft × ¼ in.) and Lubrol MO column at 100°C.[2]

much less basic ($K_b = 10^{-10}$). Mixed aryl alkyl amines have about the same basicity as aniline. Another electronegative group substituted for a hydrogen in the aryl of the aryl amine reduces the basicity of the amine even further.

Hydrogen bonding or chelation in the nitroaryl amines also decreases basicity.

When an aryl group is substituted for one hydrogen on the alkyl of an alkyl amine, the amine behaves like an alkyl amine. Therefore, benzylamine (phenylmethylamine) is water-soluble. Toluidines, which are methylanilines, are even less soluble in water than aniline. Toluidines resonate and thus reduce the basicity even more. Two aryl groups reduce the basicity almost to neutral. The diaryl amines and diaryl alkyl amines plus $XArNH_2$ (negatively substituted amines) do not dissolve in 6 N hydrochloric acid. Triaryl amines are neutral.

One acyl radical has the electron-attracting power of three aryl groups. Aliphatic amides are neutral, like the triaryl amines. The aromatic amides are very slightly acidic, which makes them more soluble in sodium hydroxide solution (5%). The sulfonamides are distinctly acidic, dissolving to a certain extent in sodium bicarbonate solution (5%). The imides, having two acyl or aroyl groups replacing the hydrogen in ammonia, are acidic. Dibasic acids are the imides most commonly encountered.

N-Substituted amides are of the mixed type, which have either aroyl or acyl groups and alkyl or aryl groups replacing the hydrogens in ammonia. Compounds of this class are either neutral or acidic, depending upon whether the specific groups are weakly or strongly electronegative. Substances in this class ($RCONH_2$ and $RCONHR'$) are probably hydrogen-bonded between their own molecules because they have higher melting points and lower solubilities than $RCONR'_2$.

Low-molecular-weight primary amines are soluble in water up to n-butylamine. The secondary amines are less soluble in water than the primary amines, and the tertiary amines are least soluble. Branching also decreases the solubility.

When water is the test solvent, amines dissolve by forming ions with the water:

$$R_3N + HOH \rightleftharpoons R_3NHOH \rightleftharpoons R_3N\!:\!H^+ + OH^-$$

When the amine substitution is too large to allow easy ionization with water or renders the product insoluble in water, the amine may dissolve in aqueous hydrochloric acid (5%) due to the greater ionization and availability of hydrogen ions:

Another advantage of using gas chromatography appears in studying these solubilities: some indication of the size of the molecule may be obtained from its solubility in water, even if slight. Distributions between water and xylene may be plotted against the retention volume, as shown in Fig. 3-31, and from this plot of homologous series the size of the molecule can be approximated.

Compounds of Intermediate Basicity: NRH_2, R_2NH, R_3N, and $ArCH_2NH_2$

Lower-molecular-weight compounds are soluble in water and ether. Water-insoluble ones are soluble in hydrochloric acid. Chromatograph on basic columns or 10% Carbowax 20M treated with potassium hydroxide (3%).

Weakly Basic Compounds: $ArNH_2$, $ArNHR$, $ArNR_2$, $RCONR_2$

These compounds are usually insoluble in water and soluble in hydrochloric acid. Chromatograph on basic columns, such as Chromosorb 103.

Essentially Neutral Compounds: Ar_2NH, Ar_2NR, Ar_3N, $RCONH_2$, $RCONHR$, $ArCOHN_2$

These are usually insoluble in water and hydrochloric acid. Chromatograph on neutral columns, such as SE-30 (silicone gum rubber).

Weakly Acidic Compounds: $ArNHR$, $(RCO)_2NH$, $ArCONHR$, RSO_2NH_2, RSO_2NHR, $ArSO_2NH_2$, $ArSO_2NHR$

Some are soluble in water. As the electronegative qualities increase, the solubility in sodium hydroxide solution increases. They may need to be chromatographed on neutral (SE-30) silicone gum rubber or acidic columns.

TYPES OF SOLUTION

Physical Solutions

Solutions may be formed by physical rather than chemical forces if two principles are considered: (1) they approximately obey Raoult's law or (2) solvent and solute have dipole-dipole inner attractive forces.

Regular Solutions. These involve solvents and solutes that do not have hydrogen-bonding tendencies, do not have polar qualities, lack solvating tendencies, and do not have solvent-solute interreactivities. Solutions in benzene or carbon tetrachloride belong in this group, as do solutions of chromatographic components on completely halogenated column coatings, e.g., chlorowax.

Intermediate Solutions. Dissolution of a solute by a solvent is a complex physicochemical phenomenon. Hildebrand[16] has pointed out that several types of physical and chemical forces often exert their independent powers of retarding or inducing solution. The purely physical dilution or dispersion of the solute into the solvent is characterized as the simplest solution. When there is a definite chemical reaction of the solvent with the solute, e.g., the reaction of water-insoluble acid with an aqueous sodium hydroxide solution to produce a water-soluble salt, there are also a great variety of intermediate phenomena that cause the solubility of a compound.

Hydrogen Bonding. The mutual attraction of two electronegative atoms for a hydrogen proton produces hydrogen bonding. Only F, O, N, and Cl form hydrogen bonds, decreasing in activity from F to Cl. Water is a strongly hydrogen-bondable solvent. Nonelectrolytes do not dissolve in water unless they can form hydrogen bonds with the water. The solubility of a given compound in water depends upon how many hydrogen-bondable groups are present in the molecules and the size of the hydrocarbon portion of the molecule. The retention of compounds on chromatographic columns also depends upon hydrogen bonding between the compounds and the liquid coating on the column. For example, the hydrocarbons ethane to butane and benzene, plus their halogen derivatives, are only very slightly soluble in water because there are only small amounts of hydrogen bonding between the hydrocarbon and water or halogen and water. Introduction of hydroxy or amine groups into these molecules causes them to become much more soluble due to the increase in hydrogen bonds. The hydrogen bond is about 5 percent of the covalent H—O bond (110 kcal vs. 5 kcal) or the covalent H—H bond (103 kcal).

In ether and nonaqueous solvents such as alcohols, and sometimes chloroform, the observable solubility is actually produced by hydrogen bonding. Many compounds that do not dissolve in carbon tetrachloride will dissolve in chloroform because it contains a hydrogen capable of donating to the hydrogen bond. For example, as shown in Fig. 3-33, when alcohols are chromatographed on a paraffin column (C or O), the branched alcohols do not readily separate from the normal alcohols. With a highly polar column such as Carbowax (K), the alcohols are retained longer (hydrogen-bondable) and thus separate more readily.

Chelation. This special case of hydrogen bonding occurs when bondable groups are located in resonating molecules (aromatic, etc.) in positions permitting hydrogen-bond formation to take place, thus forming cyclic groups. Chelation has a dramatic effect on solubility, retention on gas-chromatographic columns, and related physicochemical phenomena. When electron-attracting groups like —HC=O

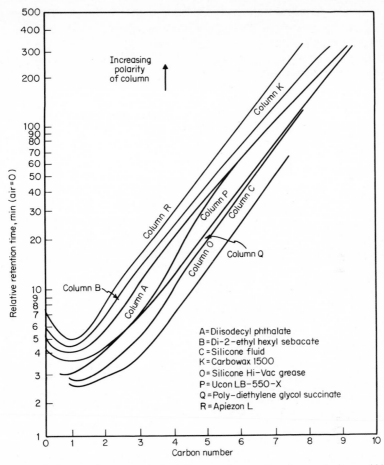

Fig. 3-33 Behavior of water and alcohols on columns of various polarity at 100°C (20 percent liquid phase, 4 m × ¼ in.).

or —NO₂ are positioned ortho to the —OH group in a phenol (or to the —NH₂ group in an aniline), the compounds have unusual properties in regard to their solubility and chemical reactions. The hydrazines behave similarly:

o-Salicylaldehyde   *o*-Nitrophenol   *o*-Nitrophenyl-hydrazine

Hydrogen bonding internally inactivates the ability of the adjacent group to hydrogen-bond with water, form salts with sodium hydroxide, or take part in other characteristic reactions of the phenol. This hydrogen bonding also prevents formation of association polymers between molecules of the compound. The behavior of these substances on gas-chromatographic columns differs from that of the meta or para isomers (see Fig. 3-34).

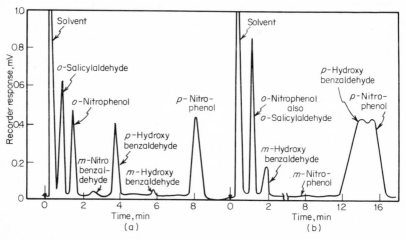

Fig. 3-34 Behavior of *o*-, *m*-, and *p*-nitrophenols and hydroxybenzaldehydes on (*a*) nonpolar 10% silicone W-98 (6 ft × ⅛ in. at 150°C, 100 ml/min He flow) and (*b*) polar columns (5% Carbowax 20M, 3 ft × ¼ in., at 225°C, 100 ml/min He flow). (*Note:* The long retention time of para isomers is due to hydrogen bonding, especially with the polar column; orthos bond with themselves.)

Association Polymers (Hydrogen-bonded). Probably one of the most marked effects on solubility is the affinity of certain molecules for hydrogen bonding together as *association polymers*. Thus, if the molecules of the solvent associate, the probability that the solute will diffuse into the solvent will be reduced. Similarly, if solution is assisted by hydrogen bonding of the solute to the solvent, association reduces the solvent-solute interaction. As an illustration, water and low-molecular-weight alcohols are highly associated solvents; ether, benzene, carbon tetrachloride, and chloroform are unassociated solvents.

Hydrogen bonding or association of the solute molecules with each other similarly reduces the solubility in association-polymer solvents, such as alcohol and water. The reverse effect also occurs. Association may produce greater solubility in ether by reducing the degree of active polar groups present due to the association or hydrogen bonding. Since chelation of molecules does not form this type of hydrogen-bonded polymers, their molecular weights, when determined, are normal. Salicylaldehyde (*o*-hydroxybenzaldehyde) is more soluble in ether than its isomer, *p*-hydroxybenzaldehyde, because (1) its chelation has no relatively active polar group and (2) it has a lower molecular weight than its association-polymer isomer.

$$HO-\langle\bigcirc\rangle-\overset{\displaystyle C=O\cdots HO}{\underset{H}{|}}-\langle\bigcirc\rangle-\overset{\displaystyle C=O\cdots}{\underset{H}{|}}$$

Even though the para isomer, which does not chelate, still has some active hydrogen-bondable groups on the terminal ends as the association polymer, it is more soluble in water than the ortho compound.

In chromatographic separation, some of these properties are useful. However, derivatives can be prepared of these compounds to minimize the effects if they become troublesome.

It can be demonstrated that many acids form similar association polymers:

$$R-C\overset{\textstyle O\cdots H-O}{\underset{\textstyle O-H\cdots O}{}}C-R$$

This association interferes with solubility in water and in alkaline solutions. It also interferes with chemical tests for specific groups, like chelation. The tests are not prevented, merely retarded.

Hydrogen bonding is reduced in chromatographic columns in both chelation and association polymers.

Oxonium Compounds. Ethers react with acids to form additive compounds of the structure $R:\overset{..}{\underset{..}{O}}:HX$, which resembles an ammonium
$\quad\quad\quad\quad\quad\quad\quad\quad\quad\quad R$

salt $H:\overset{\overset{\textstyle H}{..}}{\underset{..}{N}}:HX$. The compound would be called an oxonium salt.
$\quad\quad\quad\quad H$

The H_3O^+ ion (R_2HO^+) has been designated the oxonium ion because of its resemblance to NH_4^+, the ammonium ion. Ethers dissolve in concentrated sulfuric acid and separate out again when the acid is diluted with water. It is highly probably that the ethers react with the acid to form oxonium salts. The solubility of aldehydes, ketones, and esters in concentrated sulfuric acid is probably due to the formation of oxonium salts. Hydrogen ion in water that has been hydrated is known as *hydronium ion*, H_3O^+.

Oxonium compounds do not chromatograph but decompose to the ether and the acid (or SO_3 and H_2O).

Chemical Solutions

These types of classical chemical solutions are usually characterized as solubility caused by direct reaction between the solute and solvent. Chemical equations can be written for the reactions.   Thus, unsaturated linear or branched hydrocarbons dissolve in concentrated sulfuric acid by reaction to form alkyl sulfuric acids.   Certain aromatic hydrocarbons also react with concentrated sulfuric acid as well as alkyl alcohols. However, the alcohols dissolve due to dehydration to form alkyl sulfuric acids or dialkyl sulfates and to their tendency to form oxonium compounds.

Two major types of compounds are classified by their chemical reactivity: water-insoluble compounds that dissolve (1) in dilute acids and (2) in dilute bases.

Salts of acids or bases do not chromatograph unless they are made into volatile compounds, such as a sodium salt converted into a volatile methyl ester, etc.

Solubility by Bases. Bicarbonate- and hydroxyl-ion bases in aqueous solutions are the reagents used to classify acidic substances:

$$C_6H_5COOH + HCO_3^- \longrightarrow C_6H_5COO^- + H_2CO_3$$
$$C_6H_5OH + OH^- \longrightarrow C_6H_5O^- + HOH$$

Sodium bicarbonate and sodium hydroxide are the reagents used to furnish these negative ions.   Sodium is the only positive ion.   Thus, the hydroxyl ion is stronger than the bicarbonate.   Acids that dissolve in sodium hydroxide may be weaker acids than those which dissolve in sodium bicarbonate.   These acids form polar salts that are soluble in water.   The solution of an organic acid in a solution of sodium hydroxide involves chemical conversion of a less polar compound into highly hydrated more polar ions.   Water-insoluble carboxylic, sulfonic, sulfinic acids, and more acidic phenols are soluble in sodium bicarbonate as well as sodium hydroxide solutions.   The phenols, thiophenols, aryl sulfonamides of primary amines, certain enols, certain imides, oximes, primary and secondary nitro paraffins, and N-substituted hydroxylamines are soluble in sodium hydroxide but not in a solution of sodium bicarbonate.   Compounds which contain the $=N-OH$ group are soluble in sodium hydroxide: nitro paraffins convert to the aci isomer, which explains their solubility:

$$CH_3N\overset{\nearrow}{\underset{O}{\diagdown}} \qquad CH_2=N\overset{\nearrow}{\underset{OH}{\diagdown}}$$

Some p-nitrosophenols have an aci isomer similar to an oxime and dissolve in sodium hydroxide.

Sodium salts of these compounds can be converted to volatile esters or volatile trimethylsilyl ethers before chromatographing.

Solubility by Acids. Short-chain amines dissolve in water due to their tendency to form ions by combining with hydrogen ions from the water.

$$RNH_2 + HOH \rightleftharpoons RNH_2 {:} HOH \rightleftharpoons RNH_2 {:} H^+ + OH^-$$

$$R_2NH + HOH \rightleftharpoons R_2NH {:} HOH \rightleftharpoons R_2NH {:} H^+ + OH^-$$

$$R_3N + HOH \rightleftharpoons R_3N {:} HOH \rightleftharpoons R_3N {:} H^+ + OH^-$$

When the substituent groups in the original NH_3 become so large or of such a nature that they decrease the solubility of the amine in water only, the amine becomes more soluble in dilute acids. This is due to the greater availability of hydrogen ions in the solutions:

$$CH_3-C_6H_4-NH_2 + H_3O^+ + X^- \longrightarrow CH_3C_6H_4NH_3^+ + HOH + X^-$$

As the length of the chain is increased or more electronegative groups are added, a decrease or reduction of the solubility in both water and acid results:

Alkyl < aryl < acyl < aroyl < sulfonyl

The alkyl group has about the same electron-attracting power as hydrogen. Thus, RNH_2, R_2NH, and R_3N have about the same order of basicity as NH_3 (about 10^{-3} to 10^{-5}). The aryl group is more electronegative, hence aryl amines, $ArNH_2$, are less basic (about 10^{-10}).

Hydrogen bonding or association and chelation in some nitro-substituted aryl amines decreases their basicity further. Substitution of two aryl groups reduces basicity to near neutral, making these substances insoluble even in 6 N hydrochloric acid.

Acyl groups reduce basicity equivalent to the aryl groups. Aliphatic amides are neutral; aromatic amides are slightly acidic. Sulfonamides are definitely acidic; imides without aryl or acyl groups are acidic.

Amphoteric Compounds. These compounds behave like an acid in a basic solution and like a base in an acidic solution, thanks to the presence of both acidic and basic groups. Many amino acids have amphoteric properties (especially the lower-molecular weight compounds) and form inner salts:

$$\underset{\underset{NH_2}{|}}{CH_3CHCOOH} \rightleftharpoons \underset{\underset{NH_3^+}{|}}{CH_3CHCOO^-}$$

In aqueous solution, these compounds exist as the dipolar ions. This compound and many of its homologs are soluble in water but not in ether. They may be classified in class II (or E_2).[20]

Amphoteric substances that are water-insoluble may be classified in class III-1, 2, 3 or IV-1, depending upon the basic strength of the amino group.[20]   This group establishes the extent to which the acidic portion of the molecule is neutralized in forming the inner salt.

Amino acids which are water-insoluble are usually aliphatic homologs soluble both in hydrochloric acid and sodium hydroxide solutions (but not in sodium bicarbonate).   This is true if the amino group is not substituted or has only aliphatic substituents.   Aromatic amino acids, such as phenylalanine, are usually soluble in aqueous hydrochloric acid and sodium bicarbonate solutions.   Substitution of two phenyl groups on the amino portion of the molecule, such as diphenylaminoacetic acid, produces products which are not soluble in aqueous hydrochloric acid but dissolve in sodium bicarbonate solution (especially if water-insoluble).

The aryl group attached to the nitrogen tends to diminish solubility in class III and its basic properties so that it becomes more acidic and belongs in solubility class II.[20]   When the phenyl groups are bound to the nitrogens, the compound is no longer amphoteric and belongs in solubility class IV.[1]

$$
\begin{array}{ccc}
NH_2 & NHR' & NR'_2 \\
| & | & | \\
R-CHCOOH & R-CHCOOH & R-CHCOOH
\end{array}
$$

Decreasing basicity ⟶

$$
\begin{array}{cc}
HNC_6H_5 & N(C_6H_5)_2 \\
| & | \\
CH_2COOH & CH_2COOH
\end{array}
$$

Increasing acidity ⟶

Some other amphoteric water-insoluble compounds are amino-phenols, aminothiophenols, and aminosulfonamides.   Since these compounds usually have two different functional groups, one or both must be protected for passage through the gas chromatograph.

REFERENCES

1. Bartlett, P. D., M. Roha, and R. M. Stiles: *J. Am. Chem. Soc.*, **67**: 303 (1945); **76**: 2349 (1954).
2. Brown, H. C., and B. Kanner: *J. Amer. Chem. Soc.*, **75**: 3865 (1954).
3. Burchfield, H. P., and E. E. Storrs: *Biochemical Applications of Gas Chromatography*, pp. 288–289, Academic Press Inc., New York., 1962.
4. *Ibid.*, p. 311.
5. *Ibid.*, p. 270.
6. *Ibid.*, p. 286.
7. Cheronis, N. D., and J. B. Entrikin: *Semi-micro Qualitative Organic Analysis*, pp. 109 and 205, Thomas Y. Crowell Company, New York, 1947.

8. Dal Nogare, S., and R. S. Juvet, Jr.: *Gas Liquid Chromatography*, p. 109, Interscience Publishers, Inc., New York, 1962.

9. *Ibid.*, p. 125

10. *Ibid.*, pp. 104–131.

11. Gudzinowicz, B. J.: *Gas Chromatographic Analysis of Drugs and Pesticides*, pp. 129, 141, Marcel Dekker, Inc., New York, 1967.

12. Hammett, L. P.: *Physical Organic Chemistry*, p. 38, McGraw-Hill Book Company, New York, 1940.

13. *Ibid.*, p. 45.

14. Hammond, G. S., and D. H. Hogle: *J. Am. Chem. Soc.*, **77**: 338 (1955).

15. Harris, W. E., and H. W. Habgood: *Programmed Temperature Gas Chromatography*, p. 283, John Wiley & Sons, Inc., New York, 1966.

16. Hildebrand, J.: *Science*, **83**: 21 (1936).

17. Ingold, C. K.: *Structure and Mechanisms in Organic Chemistry*, chap. 13, Cornell University Press, Ithaca, N.Y., 1953.

18. *Ibid.*, pp. 543–550.

19. Jaffe, H.: *Chem. Rev.*, **53**: 191 (1953).

20. Kamm, O.: *Qualitative Organic Analysis*, 2d ed., pp. 8–34, John Wiley & Sons, Inc., New York, 1956.

21. Leibnitz, H. C. E., and H. G. Struppe: *Handbuch der Gas Chromatographie*, p. 578, Verlag-chemie Gmbh, Weinheim, Germany, 1967.

22. McReynolds, W. O.: *Gas Chromotographic Retention Data*, pp. 62–63, Preston Technical Abstracts Company, Evanston, Ill., 1966.

23. *Ibid.*, p. 144.

24. *Ibid.*, p. 145.

25. *Ibid.*, pp. 298–299.

26. *Ibid.*, pp. 190–191.

27. *Ibid.*, p. 216.

28. *Ibid.*, p. 147.

29. Othmer, D. F.: *Encyclopedia of Chemical Technology*, vol. I, Interscience Publishers, Inc., New York, 1961.

30. Purnell, H.: *Gas Chromatography*, John Wiley & Sons, Inc., New York, 1962.

31. Roberts, J. D., and W. T. Moreland, Jr.: *J. Am. Chem. Soc.*, **75**: 2167 (1953).

32. Schneider, F. L.: *Qualitative Organic Micro-analysis*, p. 146, Academic Press Inc., New York, 1964.

33. Seidell, A.: *Solubility of Organic Compounds in Water*, vol. 3, D. Van Nostrand Company, Inc., New York, 1940.

34. *Ibid.*, vol. 2.

35. *Ibid.*, vol. 1.

36. Shriner, R. C., R. C. Fuson, and N. Y. Curtin: *Systematic Identification of Organic Compounds*, 4th ed., p. 69, John Wiley & Sons, Inc., New York, 1959.

37. *Ibid.*, p. 64.

38. *Ibid.*, pp. 63–91.

39. *Ibid.*, p. 71.

40. *Ibid.*, p. 66.

41. *Ibid.*, p. 69.

42. *Ibid.*, p. 73.

43. Stillson, G. H., D. W. Sawyer, and C. K. Hunt: *J. Am. Chem. Soc.*, **67**: 303 (1945).

44. Szymanski, H.: *Lectures on Gas Chromatography*, Plenum Press, New York, 1963.

45. Taft, R. W., Jr.: *J. Am. Chem. Soc.*, **75**: 4731 (1953); **76**: 305 (1954).
46. Weast, R. C.: *Handbook of Chemistry and Physics*, 45th ed., pp. C-228, B-386, C-547, 313, Chemical Rubber Publishing Company, Cleveland, Ohio, 1965.
47. Wheland, G. W., *Resonance in Organic Chemistry*, John Wiley & Sons, Inc., New York, 1965.

CHAPTER FOUR

Measurement
of Physical Properties

The measurement and determination of physical properties can be of considerable assistance in identifying or characterizing an unknown compound. The techniques to be used and the degree of accuracy required depend somewhat upon (1) the type of sample, (2) the amount available, and (3) the degree of purity. Where unlimited or substantial amounts of sample of a reasonable degree of purity are available, macro methods are preferred. Even if the sample requires a good deal of purification before the physical properties are determined, macro methods are still best unless the purification procedures result in small volumes of pure product. In most instances methods have been presented to utilize large, medium, or small samples. The degree of accuracy is, in a small measure, dependent upon the sample size, although some expert microanalysts can determine physical properties with an accuracy comparable to that of macro techniques.

Occasionally, the sample itself will dictate the technique required to measure its physical properties. For example, with very reactive samples, such as halogenated silanes, the measurements must be made in a dry bag or dry box in an inert atmosphere. The size of sample will then depend upon the size dry bag or box available in which the determination can be made and the equipment enclosed.

Good physical measurements can sometimes give excellent leads to the identity of a compound. For example, if the liquid density is above 1.0, the analyst can be reasonably certain that the sample is not a normal hydrocarbon, as most hydrocarbons have densities below 1.0. The boiling point will usually tell whether it is a pure or impure compound or a mixture.

Gas-chromatographic examination will reveal the purities of these compounds as well as other properties, such as functional groups.

MELTING POINT

The melting point is defined as the range of temperatures over which the solid phase of the sample changes to liquid. In samples separated by gas chromatography, frequently only minute quantities are available. The sample can be collected on a cold microscope cover glass and the

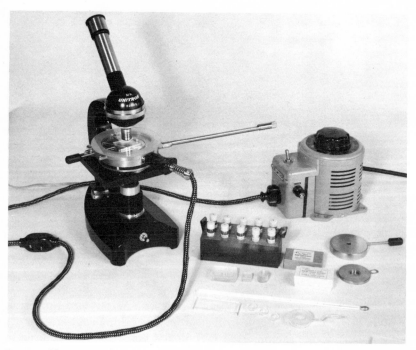

Fig. 4-1 Hot-stage microscope for determining melting point. (*A. H. Thomas Co., Philadelphia.*)

melting point determined on the hot stage of a microscope, as shown in Fig. 4-1.

If sufficient sample is available, the melting point can be determined on a Fisher-Johns melting-point apparatus or in a capillary tube using a Thomas-Hoover melting-point apparatus.

Some investigators have collected chromatographic fractions in capillary tubes arranged as shown in Fig. 4-2. The capillary tip is passed through a silicone rubber septum and pressed against the exit port of the instrument as the peak is emerging. If the sample does not condense in the tube, the remainder of the capillary is passed through a paper cup and the cup filled with Dry Ice. When sufficient material has condensed, one end is sealed and the contents melted down to the bottom. After being attached to a thermometer with a rubber band, the sample is run by the capillary method. Other physical measurements may also be made, such as density, refractive index, etc.

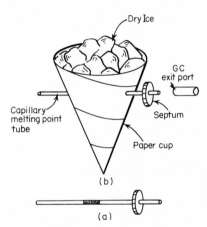

Fig. 4-2 Equipment for collection of gas-chromatographic peaks in capillary tubes: (*a*) uncooled and (*b*) cooled.

Mixed Melting Points

Often a quick identity can be established for a compound if the compound is mixed with another known compound of identical melting point and of similar properties. (If its retention time on the same column is identical to that of the known, it probably will have the same melting point.) A mixture is prepared of the two compounds, known and unknown, both of whose melting points have been determined independently. Upon melting the mixture, if both are identical, there should be no change in melting-point behavior. However, this alone cannot be used for positive evidence of the identity of the unknown.

Mixtures of two nonidentical samples show a definite melting-point depression[30] even though the melting points may be identical individually.

A few pairs of substances show no mixed melting-point depression, but usually the inability to depress the melting point is observed at only certain compositions. This can be verified by making 80/20, 50/50, and 20/80 percent mixtures and running their mixed melting points.

Capillary-Tube Method

Use a thin-walled capillary tube of about 6 cm × 1 mm diameter sealed at one end and fill it with a small amount of the powdered sample. Fasten the tube to a thermometer with a rubber band and insert it in a clamp in a beaker of colorless silicone, mineral, or vegetable oil. Heat slowly at a rate of 1 to 2°C/min, stirring the oil continuously.

Watch the sample carefully and note when it starts to melt and when it is completely liquid. Record these temperatures as the melting-point range. It is assumed that the thermometer is correct or has been corrected by using known pure materials. Table 4-1 shows values for some known pure materials, which may also be run simultaneously.

TABLE 4-1   Standards for Melting Point

Compound	M.P., °C	Compound	M.P., °C
Naphthalene	80	Fumaric acid (sublimes, 200)	287
o-Methoxybenzoic acid	100	Saccharin	229
Benzoic acid	122	Caffeine	238
Salicylamide	140	Pentaerythritol	260
Adipic acid	152	Theobromine	357
Ergosterol	168		
Thiourea	177		

SOURCE: *The Merck Index*, 8th ed., Merck and Co., Rahway, N.J., 1968.

With gas chromatography, the melting point can give the investigator an indication of the actual or relative retention time on a given column, as shown in Fig. 4-3.

Conversely, the melting point can be predicted from the chromatographic retention data using plots of these data versus their corresponding melting points. However, melting points are not as reliable a property as boiling points, especially for the lower members of a homologous series.

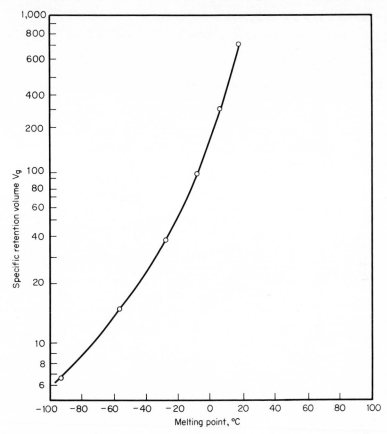

Fig. 4-3 Relationship of the melting points of *n*-alkyl hydrocarbons to their specific retention volume on an SE-30 column at 120°C.[22,35]

Hot-Stage Method

Place the compound between two 18-mm microscope cover glasses and insert on the hot stage of the Fisher-Johns apparatus or under a microscope equipped with a hot stage. Using the eyepiece on the hot stage or the microscope eyepiece, watch the sample carefully as it is heated. The temperature is read when the crystals begin to melt and again when the crystals have completely disintegrated.

Check the instruments for accuracy by running a range of known standard compounds (see Table 4-1).

Observing the crystals under a microscope through crossed polarizers makes it easier to determine the melting point as the colors disappear when the crystals melt. Also changes in crystalline form can be observed upon heating under the polarizer.

Precautions in Melting-Point Determinations

The melting point is a more difficult property to predict by chromatographic retention data, especially for the lower compounds of a homologous series.   For more accurate data, it is essential to use a chromatographic column that has very little polar influence on the sample. For example, the SE-30 column can be used, as shown in Fig. 4-4.   Even then there may be some discrepancies in the relationship of the melting point to the retention data.   In Fig. 4-4 the relationship of the odd-numbered fatty acids is different from that of the even-numbered ones.   This is probably due to a different association in the solid versus

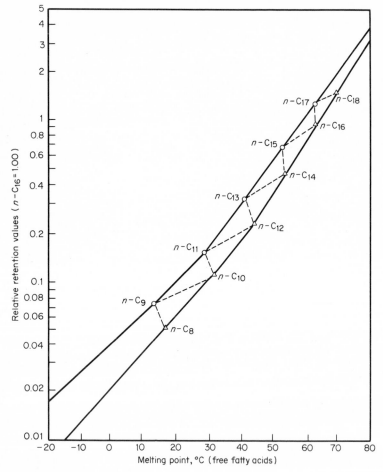

Fig. 4-4   Relationship of the melting points of the odd- and even-numbered fatty acids vs. the relative retention values of their methyl esters on a silicone grease column at 200°C.[7,32]

liquid states. This principle also holds for the fatty alcohols (see Fig. 4-5).

The melting point is a satisfactory means of evaluating the purity of a compound. A sharp melting point (range of 1 to 2°C or less) usually indicates a high degree of purity. The greater the range over which the compound melts, the less pure it is. Chromatographic examination of the compound usually will confirm the degree of purity.

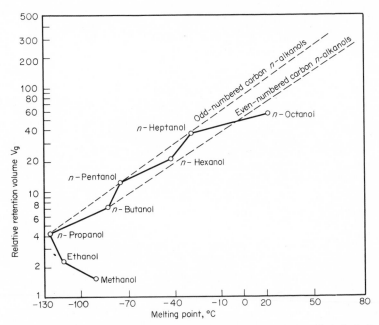

Fig. 4-5 Behavior of melting points of odd- and even-numbered n-alkyl alcohols vs. their specific retention volumes on SE-30 column (20%) at 160°C.[22, 35]

The hot-stage melting point under the microscope will reveal additional properties, especially with a polarizer, e.g., crystal changes dehydration of crystals, sublimation, and transitions from various forms. Fatty acids or alcohols can be observed under crossed polarizers and the melting point measured accurately. The disappearance of the polarized pattern indicates transition from crystal to liquid. Any chemical changes during melting can similarly be observed.

FREEZING-POINT MEASUREMENTS

Place a few milliliters of the unknown sample (liquid) at room temperature in a small test tube, which is placed in a larger tube surrounded with a cooling mixture (ice, ice plus salt, Dry Ice and acetone or liquid nitrogen). Stir the sample vigorously until it starts to freeze; remove; and continue stirring as the material solidifies. The steady-state temperature reached as the substance solidifies is considered the freezing point. The freezing point can also be observed under a cold-plate microscope.

Some versatile chemists have equipped a thermoelectric cooler with a magnifier.

REFRACTIVE INDEX

The absolute refractive index of a compound is the ratio of the velocity of light in a vacuum to that in the compound under observation. Since it is quite difficult to determine refractive index under vacuum conditions, measurement in air is commonly used as the basis of comparison. The index of refraction n carries a superscript number indicating the temperature at which the measurement is made and a subscript letter (or number) designating the wavelength of light used. The D line of sodium (589 nm) is usually the standard wavelength of light used for the observation. The refractive index of water is $n_D^{20} = 1.3330$ and $n_D^{50} = 1.3290$. Note that the refractive index decreases as the temperature is raised. This change with temperature varies with many substances, but as an approximation it may be taken as 0.0004 per degree Celsius.

The refractive index is one of the more important physical constants and can be accurately measured. It is more reliable than the boiling point for determining the purity of liquids. It is useful for the identification of an unknown organic compound. If the compound is highly pure, the refractive index helps exclude certain substances, thus narrowing the identification. In reference to chromatographic fractions, the relative retention volume V_g can be plotted against the refractive indices for a given homologous series (as in Figs. 4-6 and 4-7) and may be useful in indicating the indentity of the unknown. These relationships hold for most homologous series.

Refractometers

Several instruments are used to measure the index of refraction, the Pulfrich, immersion, Abbe, and Nichols being the most common. The Abbe and the Nichols refractometers are the most popular and easiest to use. The Abbe requires only a drop or two of the substance and can be used at very exact temperatures by connecting the cell to a constant-temperature bath. The refractive index for the sodium D line is read from the scale using the small magnifier. Results are recorded as follows:

$$n_D^{20} = 1.3330$$

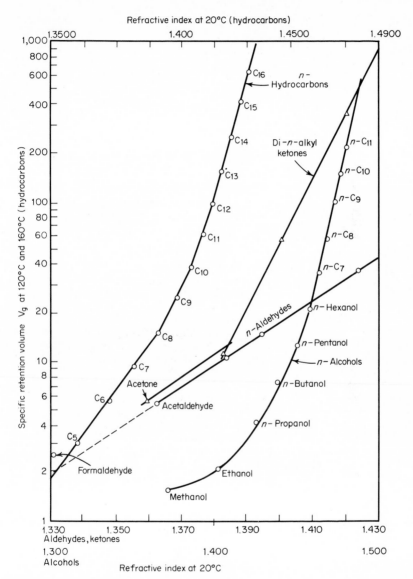

Fig. 4-6 Variation of refractive index at 20°C vs. specific retention volume on SE-30 column at 120°C (160°C for hydrocarbons) for *n*-hydrocarbons, *n*-alkyl alcohols, *n*-alkyl aldehydes, and di(*n*-alkyl) ketones. [22,35]

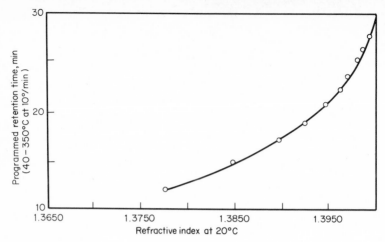

Fig. 4-7   Linear siloxanes: behavior of the programmed retention time on SE-30 column (20%) (6 ft × ⅛ in.) from 40 to 350° at 10°/min vs. the refractive index at 20°C.[18, 22]

TABLE 4-2   Refractive Index of Some Liquids[35]

Substance	n_D^{20}	Substance	n_D^{20}
Methyl alcohol	1.3288	Ethyl iodide	1.5138
Water	1.3330	Chlorobenzene	1.5248
Acetone	1.3588	Ethylene bromide	1.5379
Ethyl acetate	1.3721	o-Nitrotoluene	1.5466
n-Heptane	1.3876	Nitrobenzene	1.5562
n-Butyl alcohol	1.3991	o-Toluidine	1.5688
Methylcyclohexane	1.4253	Quinoline	1.6268
Ethylene chloride	1.4448	Aniline	1.5864
Cyclohexanol	1.4658	o-Iodotoluene	1.6095
Triethanolamine	1.4852	s-Tetrabromoethane	1.6277
Toluene	1.4961	α-Bromonaphthalene	1.6585

The instrument is calibrated by checking a liquid of known refractive index (as shown in Table 4-2) approximately the same as that of the unknown sample.

Micro or Semimicro Refractive-Index Measurement

When only very small samples are available, i.e., less than that required to cover the prism in the Abbe refractometer (such as fractions collected from a gas chromatograph), several methods are available.

Nichols Microrefractometer.[28] This instrument consists of two cells inserted into a metal constant-temperature water jacket. The micro unit requires only a small droplet from a fine capillary micropipette. Each cell contains two matched prisms cemented to a glass disk; on the disk is engraved a fine line protected by a cover glass. To use the instrument, place the sample in the depression above the prism and cover with a cover glass. Place the entire unit on the microscope stage with the $10 \times$ ocular and $10 \times$ objective. Put a ruled micrometer disk just above the microscope diaphragm and use the light from a white frosted bulb reflecting up through the cell objective and into the ocular. When the microscope is focused, two sharp lines are observed in the field. Measure the distance between these lines by means of the micrometer disk. Read the refractive index from the calibration chart prepared for each instrument and microscope.

There are two cells with the refractometer. Use the one marked $n_D = 1.52$ for liquids with indices of refraction below 1.45 and above 1.60; use the one marked $n_D = 1.72$ for indices below 1.60 and above 1.80; that is, there is some overlap.

Calibrate each cell by filling with a pure liquid of known refractive index.

Clean cells immediately after using. Remove the liquid from the cell using a vacuum through a fine capillary tube. Wash the cell at least twice with 95% ethanol (do not use acetone, as it may affect the cement) and finally wash with dry ether. Dry the cell by blowing air through a capillary containing a plug of cotton. With ordinary light from a bulb, the accuracy can be determined only to ± 0.001. Using sodium light with good temperature control, one can obtain an accuracy of ± 0.0005.

Schlieren Method. Using one of the liquids of known refractive index (Table 4-2), fill the Schlieren cell about three-fourths full. Draw the unknown sample into a capillary tube equipped with a rubber bulb.

View the cell over a background composed of half dull black and half white paper, using a lamp in front of the cell but facing away from the observer. Have the area where the capillary enters the cell at the demarcation line between dark and white paper. Allow the sample to flow out of the capillary into the liquid in the cell. If the liquids have different refractive indices, a Schlieren line forms around the liquid entering the cell. If the shadow of the Schlieren is opposite the dark region of the background, the refractive index of the sample is higher than that in the cell. If the shadow is on the same side, the cell liquid has the higher refractive index. Repeat until the index of refraction is narrowed.

The refractive index can be estimated with a high degree of precision

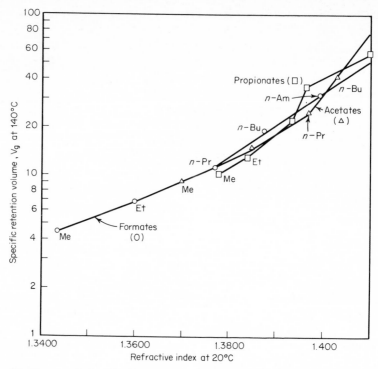

Fig. 4-8 Esters: relationship between the refractive index and the specific retention volume on neopentyl glycol succinate column (20%) at 140°C. [23,35]

by plotting the relative retention volume V_g (or relative retention time) versus the refractive index (see Fig. 4-8) of similar homologous series. Values not listed in tables can be precisely estimated in this manner.

Applications of Refractive-Index Measurements

The primary value for both density and refractive-index determinations is their use in excluding or including certain compounds in the identification techniques.   However, it is essential that the compound be pure.   If chromatographic examination or the boiling point indicates that the compound is impure, collect the heart cut (discard early and late fraction) in a preparative gas-chromatographic run or the heart cut from the distillation.

These physical constants are also very useful for verifying the identity of the compound when its probable identity has been fairly well established (see Figs. 4-9 to 4-11).   The researcher can check the structure by calculating the molecular refractivity and comparing it with the theoretical values.[11,31]

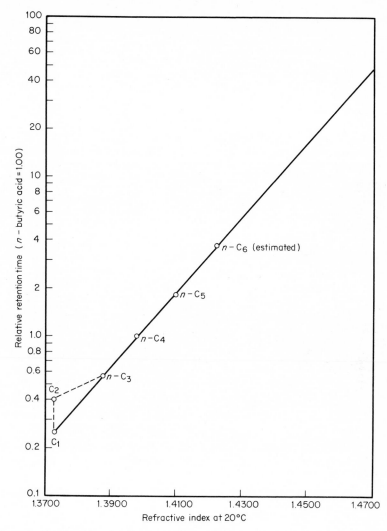

Fig. 4-9 Relationship of refractive indices of fatty acids to their relative retention time on dioctyl sebacate–sebacic acid column at 150°C.[6]

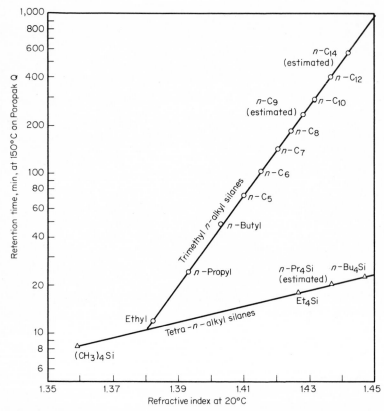

Fig. 4-10   Refractive indices of silanes vs. their retention time on Poropak QS (1½ ft × ¼ in.) at 150°C.[19]

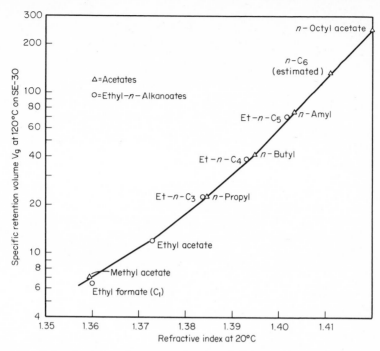

Fig. 4-11   Refractive index of esters as a function of their specific retention volume on SE-30 column (20%) at 120°C. [22,35]

In Fig. 4-12 the relationship of the molar refraction (observed) is plotted versus the relative retention volume at 120°C on an SE-30 column using gas-chromatographic techniques.   With this curve, the molar refraction for undecane was estimated to be 53.2, the value actually observed was 53.12, and the calculated value was 53.173.   Thus, gas chromatography can be a very useful tool for evaluating many physical and chemical measurements.

The Warrick molar refraction of cyclic and linear siloxanes has been plotted versus their corresponding programmed retention times on a chromatographic column of SE-30 in Fig. 4-13.   The relationships are almost linear.   The extreme values on these curves have been extrapolated from the programmed retention data to obtain fairly accurate estimated values.   This technique is quite useful where it is difficult to obtain usable values because of the extreme difficulty of providing samples of suitable purity for the actual measurements.

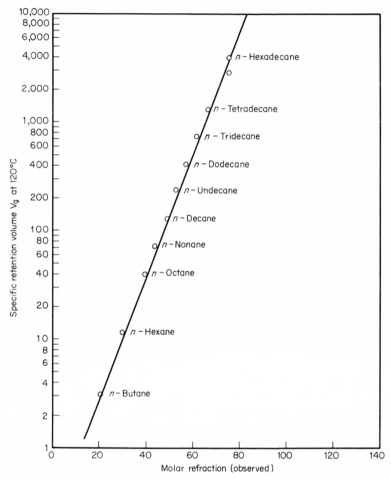

Fig. 4-12 n-Hydrocarbons: linear relationship of the molar refraction (observed) to the specific retention volume on an SE-30 column (20%) at 120°C.[2, 22]

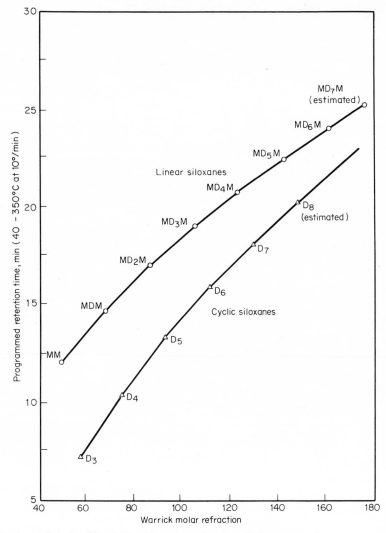

Fig. 4-13 Relationship of Warrick molar refraction of siloxanes to their programmed retention time on SE-30 (20%) (6 ft × ⅛ in.) from 40 to 350°C at 10°/min.

DENSITY DETERMINATION[34]

Density determinations assist the analyst in the identification of unknown compounds, especially those which do not form useful or well-defined derivatives. The characterization of liquid aliphatic hydrocarbons is usually verified through the determination of boiling points, refractive indices, and densities after the compounds are examined on the chromatograph.

The density determination also gives a relative measurement of the complexity of the unknown. As a general rule, compounds with a density of less than 1.0 usually do not contain more than one functional group. They may be polyfunctional if they have a density greater than 1.0.

Density measurements are not used as much as determinations of melting points, boiling points, or refractive indices because it takes a greater effort to obtain precise density values. The use of larger samples (1 to 5 ml) generally gives more accurate values if calibrated pycnometers are used and the samples are carefully equilibrated at constant temperature. Small samples (0.02 to 0.03 ml) give reliable values if suitable care is taken. Errors in weighing or losses by evaporation cause much greater discrepancies in density with the small samples.

If very limited samples are available (less than 0.02 ml) even for density measurements, the analyst can determine the relative retention volume on a chromatograph and predict the density from a plot of a homologous series, as shown in Fig. 4-14. Hydrocarbons have fairly regular relationships, but those of n-alcohols, siloxanes, and other functional compounds are not as smooth.

This is a useful technique, but the analyst must be cautioned against extending the principle too far. Extrapolation of values one or two points beyond the end of a curve may be quite valid; beyond that errors may increase.

Macro Technique

To calibrate, clean the pycnometer (1 to 2 ml capacity), dry, and weigh. Fill the bulb with boiled and cooled distilled water to slightly over the mark and immerse in a water bath at 20°C. After 5 to 10 min, adjust the level in the pycnometer exactly to the calibration mark with a capillary dropper. Remove the pycnometer from the water bath, dry carefully with a chamois, and weigh. Empty the pycnometer, dry, and fill with the liquid under investigation. Adjust at 20°C as before, and dry and weigh as previously. The weight of the sample divided by the weight of water at the same temperature gives the density of the liquid. Since the density of water at 20°C is not 1.0000, a correction must be

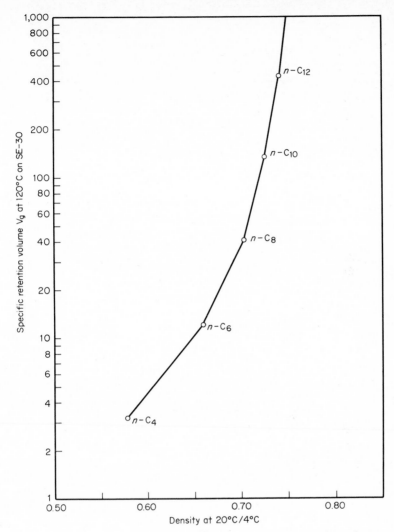

Fig. 4-14   *n*-Hydrocarbons: variation of d_4^{20} with their specific retention volume on an SE-30 column (20%) at 120°C. [22,35]

made to give the density of the substance with reference to water at 4°C as follows:

$$d_4^{20} = \frac{\text{weight of sample at 20°C}}{\text{weight of water at 20°C}} \times 0.99823$$

This is the basis of the density measurements shown in Figs. 4-15 to 4-18.

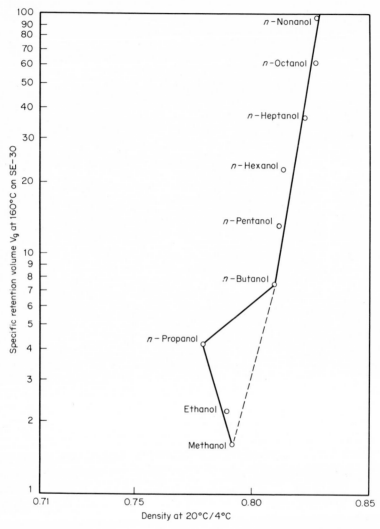

Fig. 4-15   *n*-Alcohol: variations of the density vs. the specific retention volume on an SE-30 column (20%) at 160°C. [22,35]

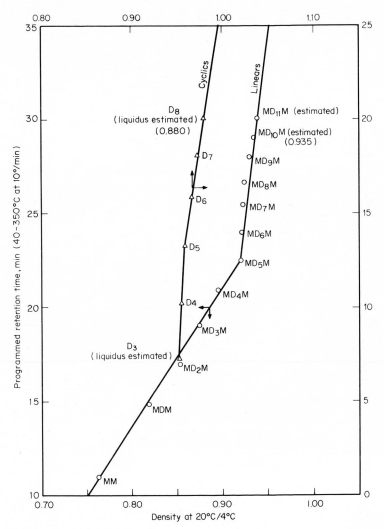

Fig. 4-16 Cyclic and linear siloxanes: variation of density with the programmed retention time on an SE-30 column (20%) (6 ft × ⅛ in.) from 40 to 350°C at 10°/min.[8, 18]

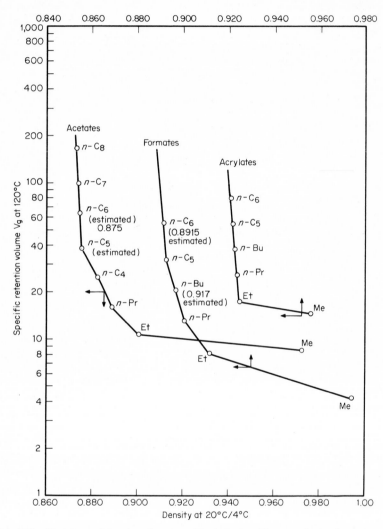

Fig. 4-17 Variation of density of esters with their specific retention volume on diethylene glycol succinate column (20%) at 120°C.[36]

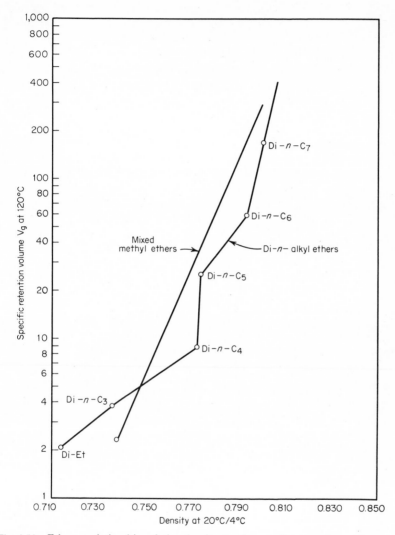

Fig. 4-18 Ethers: relationship of the density to the specific retention volume on diethylene glycol succinate column (20%) at 120°C.[36]

Micro Technique

For volumes of 0.005 to 0.016 ml, type A Alber[1] specific-gravity pipettes may be used for measurement of densities. Type B pipettes have volumes of 0.02 to 0.08 ml and a graduated scale 80 mm long divided into 1-mm divisions (bore is 0.5 mm). The pipette should be calibrated at half a dozen points on the scale using a heavy liquid selected from Table 4-3. Ground-glass caps seal both ends of the pipette, allowing the analyst to measure minute volumes of highly reactive, volatile, or hygroscopic liquids accurately. A copper wire serves to suspend the pipette on the balance.

TABLE 4-3 Some Standards Suitable for Density Measurements[35]

Compound	Density d_4^{20} °C
Pentane	0.626
Cyclopentane	0.750
Ethylbenzene	0.867
Styrene	0.909
α-Methylnaphthalene	1.002
Fluorobenzene	1.024
Ethylene chloride	1.256
Carbon tetrachloride	1.595
Iodobenzene	1.832
Ethylene bromide	2.178
Tribromopropane	2.436
Bromoform	2.890
Methylene iodide	3.325

Clean pipettes carefully with cleaning solution and rinse with distilled water and then acetone; dry in an oven prior to use. Fill the pipette from its tip by capillary action. Adjust by touching the tip against a filter paper. Cap and place in a water bath at 20°C. After 20 min remove, wipe with a dry chamois, and weigh as rapidly as possible.

Some analysts have collected small volumes of chromatographic fractions in these pipettes cooled in Dry Ice–acetone baths. The density measurements can be made directly in the calibrated pipette after suitable equilibration and adjustment. Loss of fractions is thus minimized by eliminating transfers.

Density Measurement by Gravitometer

This instrument is based on the principle that if two glass-tube manometers containing liquids of different densities are connected by means of a rubber tubing to a common source of vacuum, the heights of the liquids will be inversely proportional to their densities.   If a liquid of known density is placed in one tube, the density of the liquid in the other tube can be calculated by measuring the heights of the two liquids.   As long as the room temperature in which the measurements are made and the temperature at which the liquids are allowed to equilibrate is no more than $20 \pm 5°C$, no corrections for temperature need be made.   Most values measured by this instrument are accurate to within ± 0.1 percent.   As little as 0.3 ml of sample can be used.

Using the standard reference liquid, ethylbenzene (sp gr $d_4^{20} = 0.867$), the measurable range of specific gravities is from 0.6000 to 2.000. Carbon tetrachloride (sp gr $d_4^{20} = 1.595$) can be used for liquids of higher densities as the new reference standard.   The scale reading must be multiplied by 1.84 to use this standard.   Other reference standards can be used, as shown in Table 4-3.

Discussion

As shown in Figs. 4-15 to 4-18, the density varies with the chemical composition as well as with the structural configuration.   Density measurements can also be used to include or exclude certain probable compounds in the identification procedures.

Most hydrocarbons are identified by their physical properties, among which density is an important measurement.   As a whole, most hydrocarbons are lighter than water.   In a given homologous series (Fig. 4-19), there is a steady increase in density as the series is ascended. If these units were uniform, there would be a straight line.   However, the density decreases per increment of each $-CH_2-$ (methylene) unit. In Figs. 4-20 and 4-21, the density of the olefinic n-1-alkenes is greater than the corresponding n-alkanes.   Similarly, the n-1-alkynes are of greater density than the corresponding n-1-alkenes or n-alkanes.   Even the position which the unsaturation bond occupies in the molecule influences the density.   As the bond moves closer to the center, the density increases.   This applies to more than one set of double bonds. Thus, most liquid hydrocarbons are lighter than water.

When one hydrogen of a n-alkyl hydrocarbon is replaced with an atom of high molecular weight, the density increases, usually in proportion to the molecular weight of the atom replacing the hydrogen.   Figure 4-22 shows the relationship of the alkyl halides and their densities.   The fluorides and chlorides have densities less than that of water but greater than that of the parent hydrocarbon.   Alkyl bromides and iodides have densities greater than 1.0.   With progression up the homologous series, the densities decrease as the influence of the hydrocarbon becomes greater and that of the halide becomes less.   Eventually, all will approach the density of the parent hydrocarbon (Fig. 4-23).   Similarly,

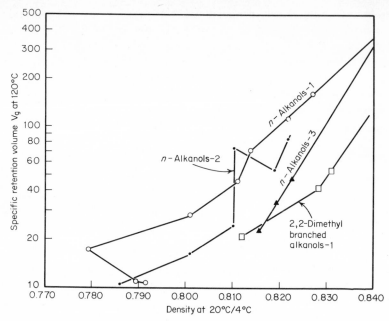

Fig. 4-19   Alcohols: relationship of the density to branching, position of hydroxyl groups, and combinations vs. specific retention volume on diethylene glycol column (20%) at 120°C.[36]

the retention times on the gas chromatograph for these halides are greater than that of the parent hydrocarbon (the first fluorides and chlorides may be exceptions).   Figures 4-24 and 4-25 show how the chromatographic retention time increases as the density increases in the following relative order (for a given homolog):

$$RH \longrightarrow RF \longrightarrow RCl \longrightarrow RBr \longrightarrow RI$$

Density increases $\longrightarrow$
Retention time increases $\longrightarrow$

By plotting similar properties on secondary, tertiary, and branched halides, similar relationships can be demonstrated.   The behavior of aryl halides follows a similar pattern, as shown in Fig. 4-26.

If the number of halogen atoms per molecule is increased, the density also increases.   Two or more of the same or different atoms also increase the density.   The higher-molecular-weight compounds of a given series usually behave more uniformly in regard to increase in density with increase in molecular weight than the lower members (see Fig. 4-27).

If functional groups are introduced, such as —OH, =CO, and —COOH, the density also increases above that of the parent hydro-

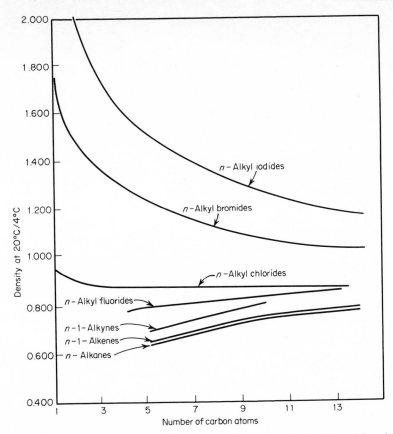

Fig. 4-20 Variation of the density of *n*-hydrocarbons and *n*-alkyl halides with carbon number.[35]

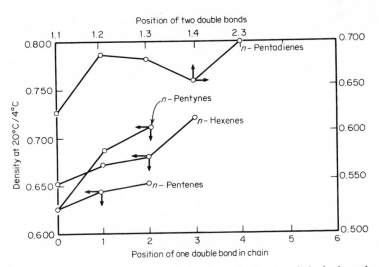

Fig. 4-21 Influence of position of double bonds on the density of the hydrocarbon.[35]

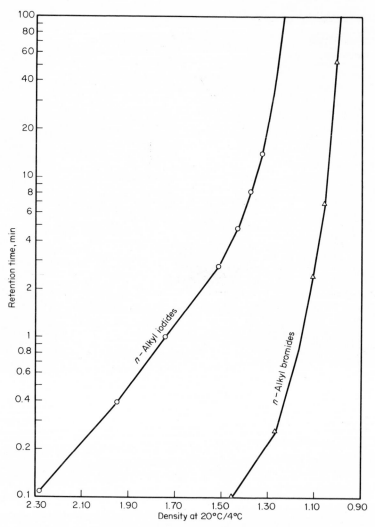

Fig. 4-22 Variation of the density of the *n*-alkyl iodides and bromides vs. their retention times on DC-200 (20%) (3 ft × ¼ in.) at 115°C.[24]

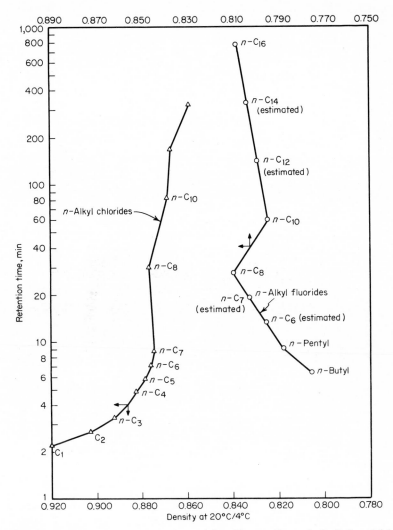

Fig. 4-23　Behavior of *n*-alkyl chlorides and fluorides on Poropak Q (3 ft × ¼ in.) at 75°C vs. d_4^{20}.[35]

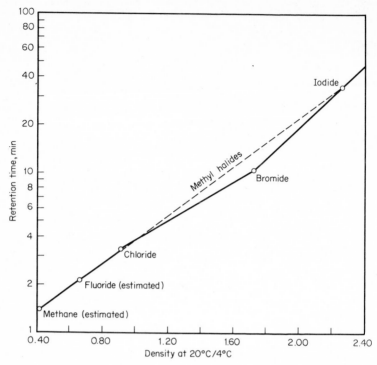

Fig. 4-24 Methyl halides: increase in density related to the increase in retention time on Poropak Q (3 ft × ¼ in.) at 75°C.[32]

carbon (see Fig. 4-28).   Introduction of the functional group increases the retention time on the chromatograph.

As noted in Fig. 4-29, the ethers are the lightest of the oxygenated substances, followed by the alcohols, aldehydes, ketones, esters, acids, and diesters.   All are less dense than water; exceptions are some of the acids and almost all the diesters.   The higher densities of formic and acetic acids are due to their association.   Ketones have density relationships similar to the alcohols and aldehydes (the curve would fall between these two).   The relationship of the ketones to retention time and density or carbon number is shown in Fig. 4-30.   One value, diethyl ketone, seems to deviate from the straight line, being denser than predicted, which may be due to some association.

The density of the alcohols becomes greater than unity if a second hydroxyl group, a halogen, or an aromatic ring is added (see Fig. 4-31). The alcohol curve shows a dip at ethanol due to its greater association than methanol.   The amines are less dense than the alcohols and show less association.

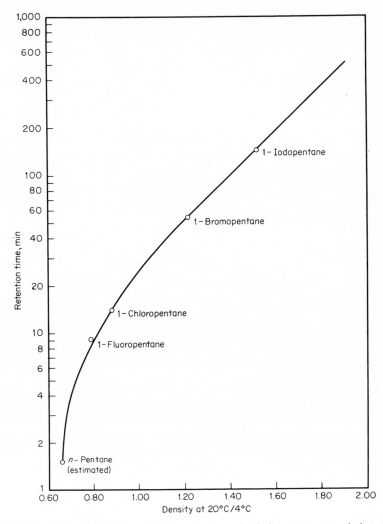

Fig. 4-25 Relation of the densities of *n*-pentane and 1-halo-*n*-pentane to their retention time on Poropak Q (3 ft × ¼ in.) at 75°C.[35]

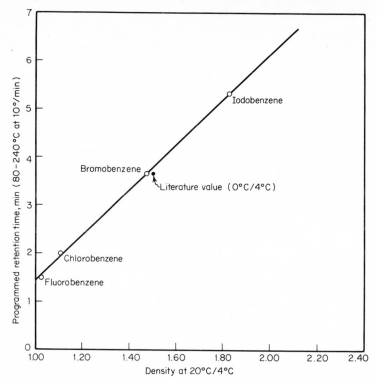

Fig. 4-26 Behavior of aryl halides (halogenated benzenes) programmed on Poropak Q (3 ft × ¼ in.) from 75–150°C at 10°/min vs. d_4^{20}.

The n-alkyl monoesters, such as the acetates, are lighter than water, but the diesters, such as n-alkyl oxalates, are heavier than water. Halogenation, hydroxylation, and ketonization of polybasic acid esters produce compounds heavier than water. If one of the monocarboxyl groups of a polybasic acid is esterified with an aromatic alcohol, a compound heavier than 1.0 is obtained. However, with increasing chain length all these compounds show a decreasing density.

A compound containing several functional groups has a density greater than that of water, especially if the functional groups tend to associate. Thus, measuring the density and noting whether it is greater or less than that of water will give a clue as to the compound's complexity.

Estimation of the density from chromatographic retention data is useful for predicting values at other temperatures. For example, if the higher and lower homologs are given densities at 20°C and the value at 30°C is desired, a simple plot of density versus retention time at 30°C (see Fig. 4-32) gives the density at the new temperature with a reasonable degree of accuracy.

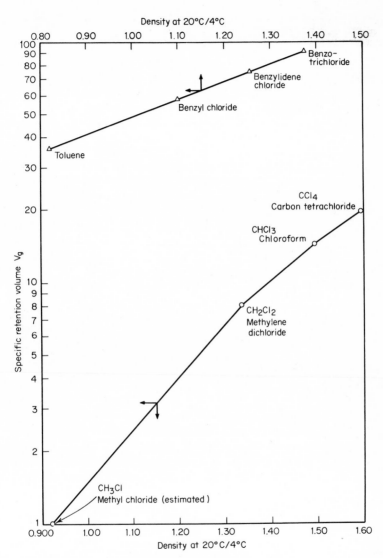

Fig. 4-27   Variation of halogenated methane and toluene densities vs. specific retention volume on SE-30 column (20%) at 120°C. [22,35]

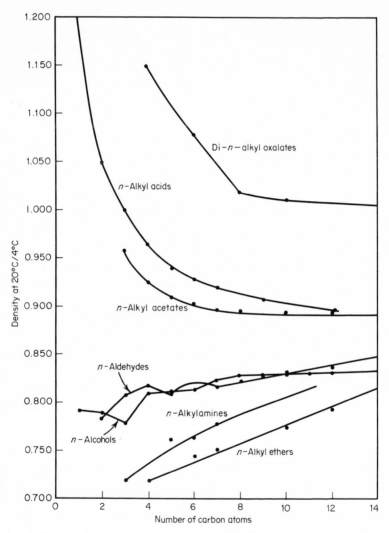

Fig. 4-28 Variation of density of homologs as the carbon number of the chain is increased.[35]

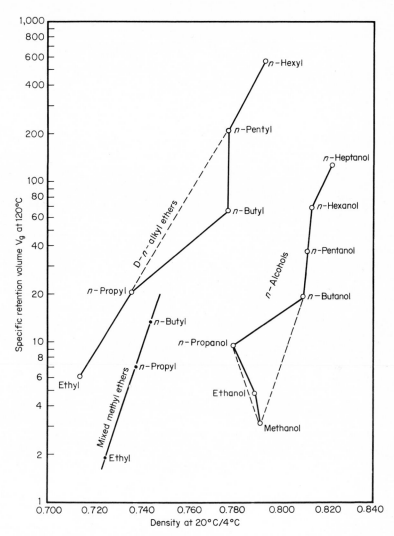

Fig. 4-29   Variation of the densities of alcohols and ethers with specific retention volume at 120°C on SE-30 column (20%).[9, 35]

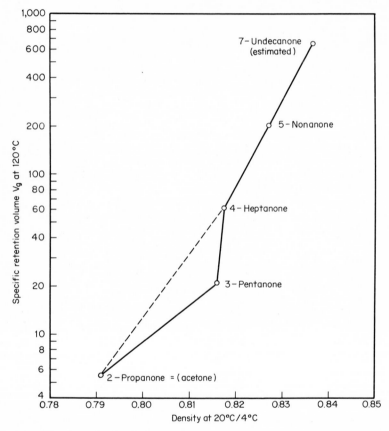

Fig. 4-30 Density of di(*n*-alkyl) ketones vs. specific retention volume on SE-30 column (20%) at 120°C. [22,35]

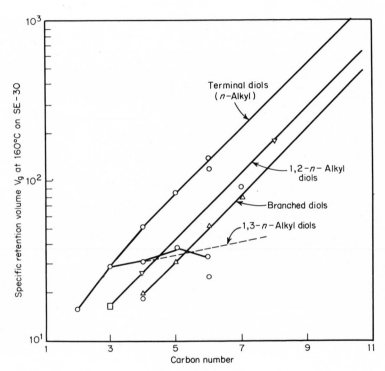

Fig. 4-31   Alkyl diols: effect of position and branching on diols vs. specific retention volume on an SE-30 column (20%) at 160°C.[22]

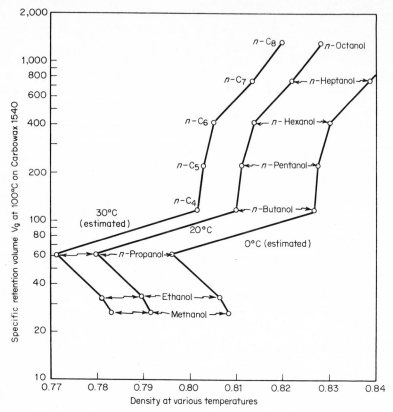

Fig. 4-32   *n*-Alcohols: variation of density with temperatures as predicted by specific retention volume on Carbowax 1540 (20%) at 100°C. [25,35]

BOILING-POINT DETERMINATION

The gas chromatograph is very useful for estimating boiling points. Columns that elute compounds by boiling point rather than by polarity are necessary for accurately predicting the boiling point of a given compound of a homologous series.   As illustrated in Fig. 4-33, hydrocarbons fall well on the straight line, but alcohols and other polar compounds, especially the lower ones in the series, tend to deviate.   For accurate prediction the retention and boiling points should be determined on components before and after the unknown in a given homologous series.

When it is necessary to verify the predicted boiling point, it can be determined by any of the following procedures.

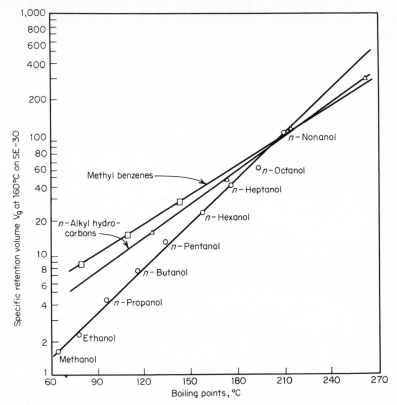

Fig. 4-33 Hydrocarbons and alcohols: relationships of boiling points to specific retention volumes at 160°C on an SE-30 column (20%).[22,35]

Macro Method

When 10 ml or more of the liquid sample is available, set up a small-scale distillation apparatus. Place a 10- to 25-ml distilling flask on an asbestos board with a 2-cm hole cut in the center. Put a small test tube in a cracked-ice bath to condense the vapors as they are distilled over. Add boiling stones to the flask to eliminate bumping. Properly place the thermometer, heat the liquid to boiling, and boil at a slow rate. Protect the flask from drafts. Discard the first 2 to 3 ml and collect the next 5 to 6 ml. Record the temperature as the boiling point and compare with the chromatographic predicted value.

The range over which the liquid boils also indicates the degree of purity of the unknown sample. If the sample boils within 1 to 2°, it is of fairly high purity. A wide range indicates that it contains considerable impurities and should be refractionated. The chromatogram should indicate more than one component. Other physical properties can be determined on the purified fractions.   Samples that are difficult to separate can be passed through a preparative gas chromatograph for a much higher degree of purification.

Determination of the boiling point on a small amount of liquid may be subject to considerable error since the vapor may be superheated or (as with high-boiling liquids) there may be substantial thermometer error.   A known compound boiling in about the same temperature range as the sample should be measured to determine the degree of correction required.

If the liquid sample is somewhat hygroscopic, it should be dried with molecular sieves and filtered to remove absorbed water.

Semimicro Method

If as much as 1 ml can be obtained for a boiling-point determination, use a Cottrell pump device.   When an equilibrium temperature is reached on the first thermometer, record the value and read the temperature on the second thermometer at the same time.   Calculate stem correction and corrected boiling point as follows:

$$BP = T_1 + N(T_1 - T_2)0.000154$$

Micro Method

Prepare a micro boiling-point tube consisting of an outer flameproof tube 5 mm in diameter $\times$ 5 cm long.   Place a capillary tube sealed in the center or about 3 to 4 mm from the end in the test tube.   Add 1 to 2 drops of the liquid sample and fasten the tube to a thermometer used for melting-point determinations.   Insert in oil bath and raise the temperature until a steady stream of bubbles comes out of the capillary and passes through the liquid.   Remove the source of heat and allow to cool slowly with continuous stirring.   The temperature of boiling is recorded at the instant bubbles cease coming out of the capillary and just before the liquid enters.   Compare with the standards shown in Table 4-4.

Boiling-Point Variations with Pressure

For accurate results, the atmospheric pressure must be noted at the time the boiling point is measured.   If the atmospheric pressure is at or near 760 mm, no correction need be made; for a pressure differential greater than 5 mm, Table 4-5 gives the degree of correction required for various boiling points for each 10-mm variation.   Usually 5 mm deviation from 760 mm is negligible.

Various investigators have proposed a variety of equations for calculating the boiling points of a wide range of organic compounds under

TABLE 4-4   Boiling-Point Reference Standards[35]

Compound	B.P., °C
Acetone	56.2
Methyl alcohol	64.96
Isopropyl alcohol	82.4
Distilled water	100.0
Chlorobenzene	132.0
Bromobenzene	156.2
Mesitylene	164.7
Iodobenzene	188.5
Ethylene glycol	197.9
Decyl alcohol	229.0
2,4-Dimethylquinoline	264–265
Glycerol	290.0
Benzyl ether	298.0
2,4,6-Trichloro-o-phenylphenol	317–318

TABLE 4-5   Temperature/Pressure Correction Factors

Correction in Celsius Degrees for Each 10 mm Difference in Pressure

Boiling point		Liquids	
°C	°K	Nonassociated*	Associated†
50	323	0.38	0.32
100	373	0.44	0.37
150	423	0.50	0.42
200	473	0.56	0.46
300	573	0.68	0.56
400	673	0.79	0.66
500	773	0.91	0.76

* For example, hydrocarbons, alkyl halides, ethers, and esters.

† For example, alcohols and acids.

vacuum.   The integral form of the Clausius-Clapeyron equation and Trouton's rule have been used but generally are not completely satisfactory.   It can be demonstrated that Trouton's constant is not the same for associated and nonassociated liquids, since there is a variation of the constant with temperature.   Weast[35] and Bordenca[5] have used empirical equations converting into graphs corresponding to eight types of compounds.

It is more convenient to determine a variety of boiling points on known compounds and develop a set of corrections to be added to the determined values in order to extrapolate to normal atmospheric pressure of 760 mm.   Germann and Knight[13] developed a set of charts for over 180 organic compounds.

To give analysts an estimate of the degree of correction required on a variety of organic liquids (nonassociated and associated) Table 4-6 gives the boiling point of a series of compounds at various pressures.

TABLE 4-6   Boiling Point of Organic Compounds at Various Pressures, °C

Compound	Pressure, mm Hg					$T_{760} - T_{550}$
	760	700	650	600	550	
n-Heptane	98	96	94	91	88	10
n-Propyl alcohol	97	95	93	91	89	8
Iodobenzene	188	185	182	179	175	13
n-Valeric acid	186	183	180	178	175	11
Fluorene..............	298	294	290	286	282	16
β-Naphthol	295	292	288	284	280	15

By extrapolation, the boiling point for the first compound should be approximately 30°C lower at 150 mm pressure than at 760 mm.   The β-naphthol boiling point should be approximately 45°C lower at 150 mm. The boiling point of an associated compound does not drop as rapidly as that of a nonassociated compound.   These values can be extrapolated from chromatographic retention data plus boiling points of other homologous series at 760 mm (see Fig. 4-34).

Applications in Predicting Properties

As shown in Fig. 4-35, the boiling points of any given homologous series increase as the molecular weight increases in a uniform pattern. However, the increase per methylene unit, $-CH_2-$, is not constant. The temperature is greater for the lower members of a homologous series and less for the higher members.   Table 4-7 gives the boiling point of the n-hydrocarbons and the increase in boiling point per methylene unit.   If one hydrogen of the n-hydrocarbon is replaced by a dissimilar atom or group of atoms, an immediate increase in the boiling point results.   This is also borne out in chromatographic data, where an increase in the retention time on a nonpolar column is found

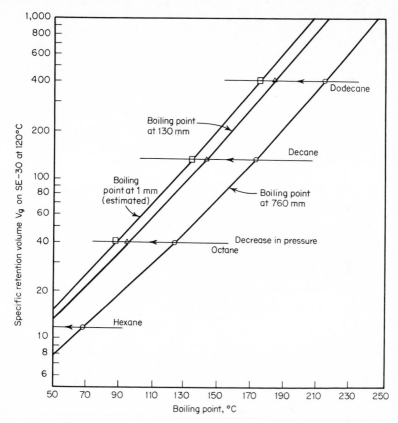

Fig. 4-34 *n*-Hydrocarbons: variation of the boiling point with pressure as predicted by plotting vs. the specific retention volume on SE-30 column (20%) at 120°C. [22,35]

similarly. Thus, *n*-alkyl alcohols, aldehydes, ketones, acids, halides, etc., show greater retention time and higher boiling points than the corresponding *n*-alkyl hydrocarbons.

If the added group promotes association, such as hydrogen bonding, a still greater increase in the boiling point is obtained. Acids and alcohols show the most distinct effect since hydrogen bonding occurs in these compounds. For example, the boiling point of *n*-pentane is 36°C and that of *n*-amyl alcohol is 138°C. The lower members show a greater effect than the higher members due to the increased influence of the associating group on the smaller hydrocarbon backbone. That is, *n*-propyl alcohol shows a boiling-point increase of 142° over the corresponding *n*-propane (see Table 4-8).

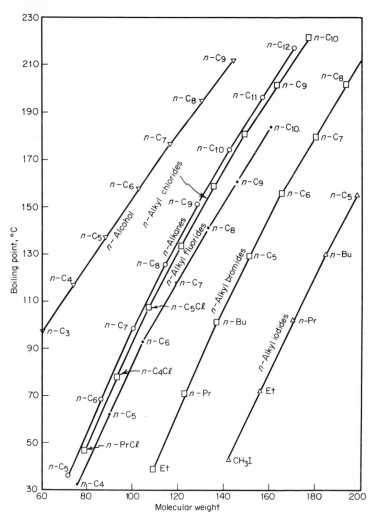

Fig. 4-35 Relationship of the molecular weight to the boiling point of some *n*-alkanes, *n*-alcohols, and *n*-alkyl halides.[32]

TABLE 4-7 **Boiling-Point Variations of n-Hydrocarbons**

Compound	B.P., °C	Increase per $-CH_2-$ unit, °C
n-Pentane..............	36.0	
		32.3
n-Hexane	68.3	
		30.1
n-Heptane	98.4	
		27.3
n-Octane..............	125.7	
		25.1
n-Nonane..............	150.8	
		23.2
n-Decane	174.0	
		22.0
n-Undecane	196.0	
		21.0
n-Dodecane	217.0	
		18.5
n-Tridecane	235.5	
		18.0
n-Tetradecane	253.5	
		17.2
n-Pentadecane	270.7	
		16.3
n-Hexadecane	287.0	

As more associating groups are added, the boiling point increases further. However, the increase is not as great as that produced by the first group. The chromatographic isothermal relative retention time (or volume) is increased logarithmically rather than arithmetically. With the proper programmed temperature retention values, linear relations are obtained (see Figs. 4-36 and 4-37). The column temperature is being increased in direct proportion to the increase in boiling point per methylene unit. The increase in boiling point per hydroxyl group is much greater than the column-temperature increase per methylene unit.

If the hydroxyl groups are converted to esters or ethers, the hydrogen-bonding tendencies are reduced. The boiling points and relative retention volumes tend to decrease similarly (see Fig. 4-38 and Table 4-9).

If the oxygen analogs are compared with their sulfur counterparts, the association by hydrogen bonding has greater influence on boiling point and retention time than molecular weight. Since the sulfur analogs do not have as high hydrogen bonding, they are only slightly

TABLE 4-8   Alcohols: Effect of Hydroxy Group on Boiling Point and Retention Time

	n-Butane		n-Butanol		1,4-Butanediol		1,2,4-Butanetriol		1,2,3,4-Butanetetrol
	CH_3 CH_2 CH_2 CH_3		CH_3 CH_2 CH_2 CH_2OH		CH_2OH CH_2 CH_2 CH_2OH		CH_2OH $CHOH$ CH_2 CH_2OH		CH_2OH $CHOH$ $CHOH$ CH_2OH
B.P., °C	−0.5		116		235		312		329
Increase per —OH unit		116.5		119		77		17	
Increase in programmed retention time per —OH unit*	0.1		1.0		6.2		6.4		6.6

	n-Propane		n-Propanol		1,3-Propylene glycol		Glycerol
	CH_3 CH_2 CH_3		CH_3 CH_2 CH_2OH		CH_2OH CH_2 CH_2OH		CH_2OH $CHOH$ CH_2OH
B.P., °C	−45		97		216		290
Increase per —OH unit		142		119		74	
Increase in programmed retention time per —OH unit*	0.1		0.8		4.8		

* On Carbowax 20M at 125 to 200°C at 4°/min.

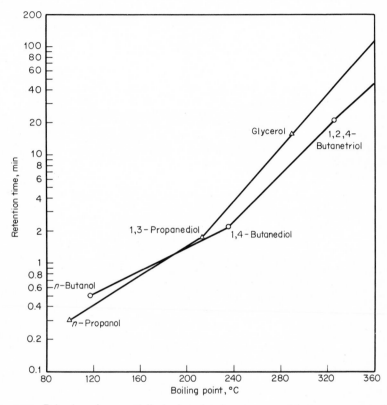

Fig. 4-36 Behavior of sequentially hydroxylated propanes and butanes on mannitol column (20%) (6 ft × ¼ in.) at 190°C vs. their boiling points.[35]

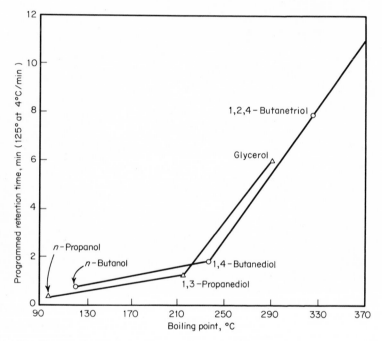

Fig. 4-37   Behavior of sequentially hydroxylated propanes and butanes on Carbowax 20M (20%) programmed from 125°C at 4°/min vs. their boiling points. [25,35]

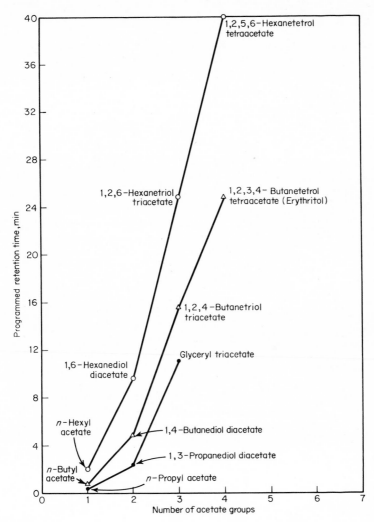

Fig. 4-38 Separation of alcohols and polyols as acetates on an XE-60 column (2%) (6 ft × ¼ in.) programmed from 100 to 250°C at 4°/min.[17]

TABLE 4-9   Effect of Ester and Ether Formation on Boiling Points and Retention Time

Esters

	n-Propanol	n-Propyl acetate	1,3-Propanediol	1,3-Propanediol monoacetate	1,3-Propanediol diacetate
	CH$_2$OH—CH$_2$—CH$_3$	CH$_2$OOCCH$_3$—CH$_2$—CH$_3$	CH$_2$OH—CH$_2$—CH$_2$OH	CH$_2$OOCCH$_3$—CH$_2$—CH$_2$OH	CH$_2$OOCCH$_3$—CH$_2$—CH$_2$OOCCH$_3$
B.P., °C	97	101.5	216	180	190
Increase in programmed retention time per —OH unit*	1.6	—†	—†	—†	9.0

Ethers

	Glycerol	Glycerol monoethyl ether	Glycerol diethyl ether	Glycerol triethyl ether
	CH$_2$OH—CHOH—CH$_2$OH	CH$_2$OC$_2$H$_5$—CHOH—CH$_2$OH	CH$_2$OC$_2$H$_5$—CHOH—CH$_2$OC$_2$H$_5$	CH$_2$OC$_2$H$_5$—CHOH$_2$H$_5$—CH$_2$OC$_2$H$_5$
B.P., °C	290	280	191	185
Decrease, °C		10	39	6
Relative retention time†	1.00	0.66	0.33	0.10
B.P. of dimethyl ethers, °C	290	220	169	148
Decrease, °C	70		49	11

* On XE 60 at 100 to 250°C at 4°/min.   † On SE-30 at 160°C.   ‡ These values unavailable.

associated.   The sulfur analogs boil lower than the oxygen analogs and have lower retention volumes per —SH unit even though the sulfur compounds have higher molecular weights.[21]

The ethers have no hydrogen attached directly to oxygen or sulfur, unlike the alcohols, acids, or aldehydes, and do not show association due to hydrogen bonding.   The alkyl sulfides boil higher than the ethers, probably due solely to their higher molecular weights; similarly, they have increased retention times (see Fig. 4-39 and Table 4-10).

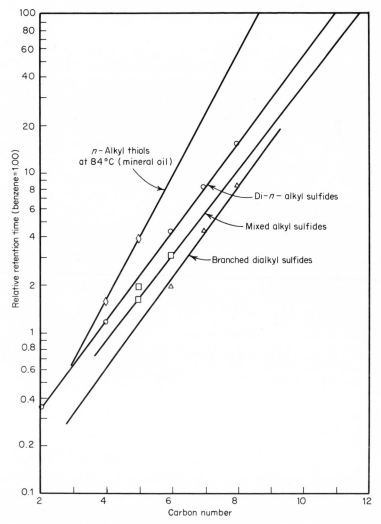

Fig. 4-39   Alkyl sulfides and thiols: effect of branching on relative retention time on silicone oil column at 100°C and mineral oil column at 84°C.[21]

TABLE 4-10   Comparison of Oxygen and Sulfur Analogs

Oxygen analog	B.P., °C	Specific retention volume, SE-30 at 120°C	Sulfur analog	B.P., °C	Specific retention volume, SE-30 at 120°C
H_2O	100	4.1	H_2S	−62	0.1
CH_3OH	66	3.1	CH_3SH	6	0.4
CH_3CH_2OH	78	4.8	CH_3CH_2SH	37	1.4
CH_3COOH	119	—	CH_3COSH	93	14.0
$(CH_3)_2O$	−24	0.1	$(CH_3)_2S$	38	10.0
$(C_2H_5)_2O$	+35	6.1	$(C_2H_5)_2S$	91	32.0
$(C_3H_7)_2O$	90	20.2	$(C_3H_7)_2S$	143	98.0
$(C_4H_9)_2O$	140	65.4	$(C_4H_9)_2S$	189	270.0

Thus, when an atom in a molecule is replaced with an atom of higher atomic weight, an increase in boiling point and retention time results as long as no increase or decrease in hydrogen bonding or association is produced by the substitution (see Fig. 4-40). The substitution of chlorine, bromine, and iodine for one or more hydrogens in a normal hydrocarbon produces successive increases in boiling point and retention time. Similarly, the substitution of oxygen, sulfur, or nitrogen increases the boiling point and retention time, but the increase may be greater than predicted due to hydrogen-bonding effects.

Effect of Branching

Branching of the hydrocarbon chain and the position of the substituent functional group (or groups) influences the boiling point as well as the chromatographic retention values. Branching has a similar effect on the solubility (see Chap. 3), as illustrated in Table 4-11 and Figs. 4-41 to 4-43.

Generalizations for Saturated Aliphatic Alcohols

1. The straight-chain n-alkyl primary alcohols have the highest boiling points and greater retention values than the branched alcohols.
2. In comparing alcohols of the same type, i.e., primary, secondary, or tertiary, the greater the degree of branching the lower the boiling points and the retention values.
3. When isomeric saturated alkyl alcohols of the same molecular

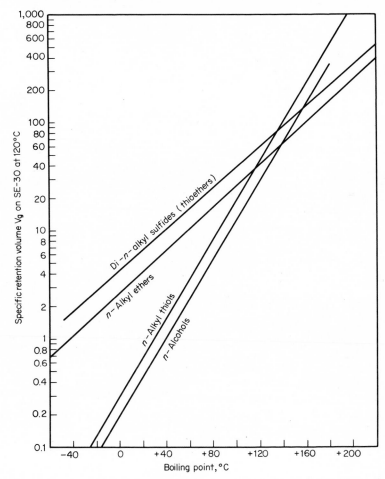

Fig. 4-40 Relationship of the *n*-alkyl ethers and alcohols to their sulfur analogs; boiling points vs. specific retention volume on SE-30 (20%) nonpolar column at 120°C.[22]

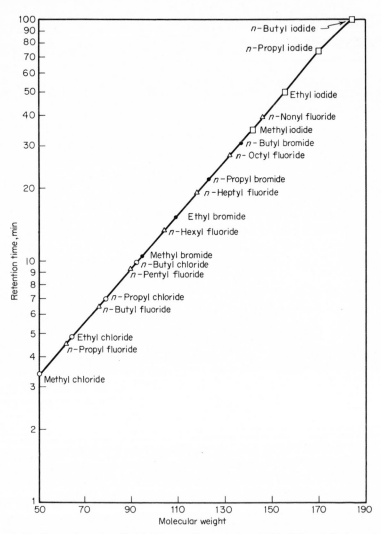

Fig. 4-41 Separation of *n*-alkyl halides on Poropak Q at 75°C (3 ft × ¼ in.) as related to molecular weight.

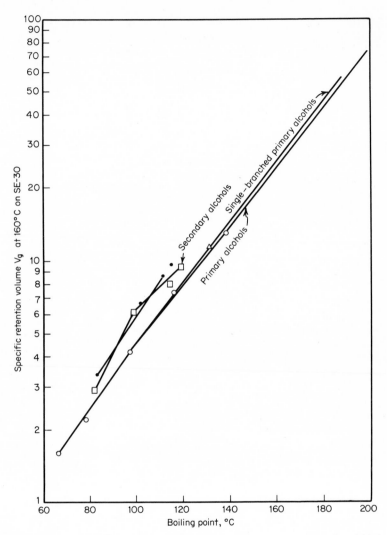

Fig. 4-42   Alcohols: separation on an SE-30 column (20%) at 160°C (6 ft × ¼ in.) vs. boiling points.[22]

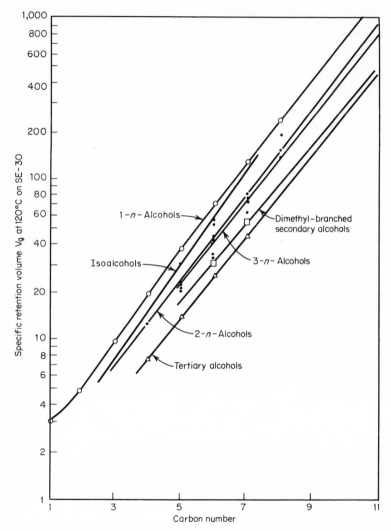

Fig. 4-43   Alcohols: effect of branching and position of hydroxyl group on specific retention volume on an SE-30 column (20%) at 120°C vs. carbon numbers.[22]

TABLE 4-11   Effect of Branching on Boiling Points and Specific Retention Volume of Alcohols

Primary	B.P., °C	Specific retention volume*	Secondary	B.P., °C	Specific retention volume*	Tertiary	B.P., °C	Specific retention volume*

CH₃OH
CH₃CH₂OH
CH₃CH₂CH₂OH
	66
	78
	97
		1.6
		2.2
		4.2
			CH₃—CH—CH₃
			       |
			      OH
				82	2.9	CH_3—C—CH₃ (CH₃, OH)	83	3.4

CH₃CH₂CH₂CH₂OH	116	7.4	CH₃CH₂CH—CH₃
			            |
			           OH
				99	6.2			

CH₃CH—CH₂OH
     |
    CH₃
	108	6.2						

CH₃CH₂CH₂CH₂CH₂OH	138	12.9	CH₃—CH₂CH₂—CH—CH₃
			                       |
			                      OH
				119	9.0			

CH₃CH—CH₂CH₂OH
     |
    CH₃
	131	11.3	CH₃CH₂—CH—CH₂CH₃
			              |
			             OH
				115	9.4			

CH₃CH₂—CH—CH₂OH
           |
          CH₃
	129	11.5						

CH₃—C—CH₂OH
     |
    CH₃ (CH₃)
	114	8.0	CH₃—CH—CH—CH₃
			     | |
			    CH₃ OH
				111	8.4	CH₃CH₂—C—CH₃ (CH₃, OH)	102	6.7

* On SE-30 (20%) column (6 ft × ¼ in.) at 120°C.

weight are compared, the primary alcohols have higher boiling points and greater retention values than the secondary alcohols. These in turn have higher boiling points and greater retention values than the tertiary alcohols.

Primary > secondary > tertiary in boiling point and retention values (decreasing boiling points and retention values → isomers of saturated alkyl alcohols of the same molecular weight)

Accurate boiling points can be predicted by chromatographic retention values, which is useful when the values are difficult to determine because of reaction, decomposition, association, or toxicity. Determination of the boiling point can assist in excluding certain types of compounds when identifying an unknown.

Generalizations for Halides

1. If an organic chloro compound has a boiling point below 132°C, it must be aliphatic in character. If the boiling point is above 132°C, it may be either aliphatic or aromatic. Chlorobenzene boils at 132°C, and 1-chlorohexane boils at 134°C. Chromatographic retention data can materially assist in verifying whether a compound is aromatic or aliphatic (see Fig. 4-44). Analogously, with a programmed retention time of 27.8 min or greater, the compound may be aromatic or aliphatic; for less than 27.8 min the unknown must be aliphatic.

2. An organic bromo compound which has a boiling point below 157°C must be aliphatic. If the unknown boils above 157°C, it may be aliphatic or aromatic. Bromobenzene boils at 157°C. Retention values will assist greatly; e.g., values under 36 min indicate aliphatic compounds; values above 36 min indicate aromatic or aliphatic (see Fig. 4-44).

3. Similarly, an organic iodo compound having a boiling point below 188°C must be aliphatic; above 188°C it can be either aromatic or aliphatic. Iodobenzene boils at 188°C. Retention values will assist in identifying the compound. When they are under 60 min, it must be aliphatic. When the retention values are above 60 min, it may be aromatic or aliphatic (see Fig. 4-44).

4. An organic fluoro compound which has a boiling point below 87° may or may not be aliphatic, since these compounds behave very differently due to the powerful electron-attraction effect of the fluorine atom. Fluorobenzene boils at 87°C; 1,4-difluorobenzene boils at 88°C; 1,2-difluorobenzene boils at 92°C. Position has a great effect on the boiling point and retention values of most halides. Fluorobenzene has a retention value under 12 min.

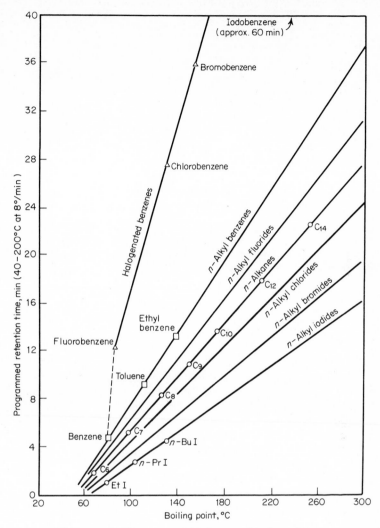

Fig. 4-44 n-Alkanes, n-alkyl benzenes, and halogenated hydrocarbons programmed on Poropak Q (3 ft × ¼ in.) from 40 to 200°C at 8°/min vs. boiling points.

STEAM DISTILLATION

Steam distillation or codistillation with water is a useful technique for separating mixtures. The layers of liquids can be separated in a separatory funnel or in a special receiver.

Saturating the aqueous phase with a salt, such as sodium chloride or potassium carbonate, will substantially reduce the overall time required

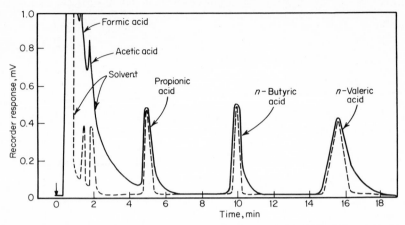

Fig. 4-45 Separation of lower fatty acids on Carbowax 20M (5%) (4 ft × ¼ in.) at 100°C with (dashed line) and without (solid line) water saturation of gas phase.[40] (*Note*: Tailing of peaks is reduced by the treatment, using flame detection which is insensitive to water.)

for the distillation and tend to increase the volatility of compounds which are normally difficult to distill.

Figure 4-45 illustrates several chromatographic separations made using an inert-gas phase saturated with water vapor simulating a steam-distillation separation.   A type of steam distillation can also be simulated by injecting a slug of water (usually twice or more the volume of the sample) in the injection needle together with the sample (see Fig. 4-46).

Steam distillation is useful for concentrating volatile and water-soluble organic compounds.   As shown in Fig. 4-47, traces of unknown

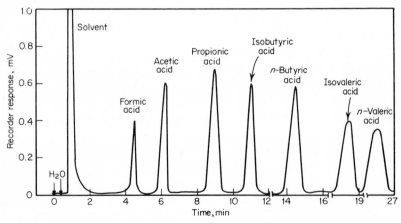

Fig. 4-46 Injection of water solution (or solvent solution plus a slug of water) of the lower fatty acids on a Tween 80 column (20%) (1 m × ¼ in.) at 110°C.

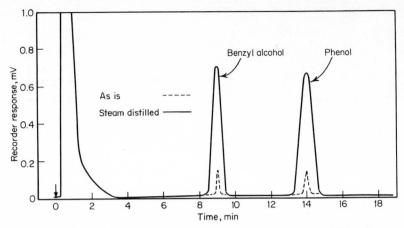

Fig. 4-47 Separation of benzyl alcohol and phenol on Carbowax 20M column at 165°C (5%) (4 ft × ¼ in.) illustrating the concentrating effect of steam distillation.

components are shown in the water sample. Steam distilling this solution concentrates the trace components and separates them from the bulk of the sample, giving more distinct peaks, which can be easily identified.

SUBLIMATION

This is the process of converting volatilizable solids to gases and condensing them back to solids without passing through the liquid state. Not all carbon compounds sublime. Those which do can facilitate separations that may be somewhat difficult otherwise. Often good crystals can be formed on microscope slides for further examination and identification.[9] Sublimation onto a slide can produce pure compounds which are then dissolved in a solvent and injected into the gas chromatograph for verification by retention data. Derivatives may also be prepared for further chromatographing.

Two types of apparatus are available for sublimation: those which can and those which cannot accurately measure the temperature at which sublimation takes place. If only one component of a mixture can be sublimed, the second type is convenient, provided that no other components in the mixture decompose or distill. The temperature of sublimation is not always a sharp fixed temperature but can be confined within definite limits by using standard equipment. The vapor pressure and the distance between the surface of evaporation and the surface of condensation are important. For substances sublimable at very low vapor pressure the distance between surfaces must be small.

Sublimation from Glass to Glass Surfaces

For a simple and rapid sublimation, clip together two watch glasses; heat the lower one and cool the upper. Two slides may similarly be used if the upper one is wide. A drop of water on the upper slide will help keep the slide cool and facilitate condensation. The condensate may be examined under the microscope, dissolved in a solvent and chromatographed, or examined by other instrumental or chemical techniques.

For larger quantities, two petri dishes may be used. One is placed with its open portion downward on an asbestos-covered wire gauze to act as an air bath. The sample is placed in another petri dish and closed with the cover. The bottom half of the second petri dish is placed on top and partially filled with water to act as a cooling medium. Schneider[30] has described a more exact temperature-controlled sublimation.

OPTICAL ROTATION

This physical property is determined only if the list of probable compounds includes optically active materials. It can assist in verifying the identity of the unknown compound or compounds. Certain chromatographic columns can be used to separate optically active compounds.[12,14,15]

Polarimetric Measurements

A popular form of polarimeter is the Lippich double-field type. To obtain the rotation of the unknown solution, place the polarimeter tube containing the solution in the trough of the instrument. Close the cover and repeat the procedure used for obtaining a blank reading. Take an average of at least five observations. The observed rotation is the numerical difference between this value and the blank value.

For exact values, the D line of sodium is usually used with electrically heated sodium-vapor lamps. The sodium flame can be used, but it is not as intense. A white lamp may be used with a suitable filter. An inexpensive filter may be made by placing 300 g of water containing 8 to 9 g of $CuSO_4 \cdot 5H_2O$ and 9.4 g of $K_2Cr_2O_7$ between the light and the instrument.

The green mercury arc line can also be used. No filter should be used between this light and the instrument.

Note that the specific rotation may be susceptible to the solvent, concentration, and other factors, including wavelength of light. The observed rotation must be run under identical conditions to those reported in the literature if valid comparisons are to be made.

Weatherall[37] has reported chromatographic separation of cis and

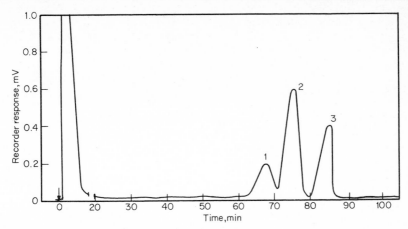

Fig. 4-48 Separation of 2,4,6-tricarbomethoxyheptane isomers on XE-60, cyanosilicone column (30%) on 60/80 mesh Chromosorb (20 ft × ⅜ in.) at 210°C.[10]

trans isomers of diols using a LAC column. Hause, Hubicki, and Hazen[17] reported the separation of isomeric polyol acetates on a silicone nitrile column (XE-60). Separation of optical isomers of polyols has been reported by various investigators on a variety of columns, including ethylene tetramine.[15]

Figure 4-48 illustrates a separation of an optically active compound by chromatography. Certain investigators have separated optically active amino acids by gas chromatography, but the chemical preparation of derivatives and subsequent chromatography must not alter the optical-rotation properties (see Fig. 4-49).

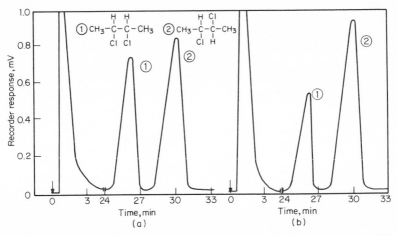

Fig. 4-49 (a) Separation of 2,3-dichlorobutane isomers on tetracyanoethoxy quadrol column (30%) on 60/80 mesh Chromosorb P at 125°C and (b) alteration of isomer ratio using higher injection-port temperature.[10]

Using careful chromatographic techniques, the analyst can obtain good separations which will correlate well with the observed optical rotations.

MOLECULAR WEIGHT

Modern methods of molecular-weight determinations are faster and more exact than the older methods. For homologous-series compounds, molecular weights can be estimated by plotting the retention time or relative retention volume versus molecular weight, as shown in Figs. 4-50 to 4-52. For nonhomologous compounds, use of the silicone column, SE-30, assists the analyst in estimating the molecular weight. Compounds are eluted by the boiling point on this column.

When it is difficult to prepare derivatives or satisfactory derivatives are not known and more exact values are required than those estimated by chromatographic values (especially if the homologous series is unknown), the Rast method is convenient.[26] The molar lowering of the melting point of camphor is quite large in comparison with other substances. The value of 39.7 for camphor[26] is valid for molar concentrations above 0.2. Solutions more dilute than 0.2 M increase the constant to approximately 50. Therefore, it is necessary to run an approximate molecular weight and then add enough solute to a second run to give a molar concentration of 0.2 to 0.5. Probably the estimate from the chromatographic plot will enable the analyst to obtain an accurate value on the first run on at least a 0.2 M concentration.

Other Methods

The vapor-phase osmometer[19] is valid up to molecular weights of about 10,000. The membrane osmometer[3] measures molecular weights above 10,000. The gas chromatograph has determined molecular weights above 2,500 with heat-stable substances, such as siloxanes.[8] Use of silicone derivatives, e.g., trimethylsilyls, has stabilized heat-sensitive substances of fairly high molecular weight such as di- and trisaccharides.

Boiling-point elevation[27, 32, 33] has been used but requires considerable time to attain equilibrium with the solvent and after addition of the unknown compound. High-molecular-weight compounds have been analyzed by light scattering.[38]

Compounds which have very high molecular weights, which are not volatile enough, and which cannot be made volatile by derivatization can be degraded in a pyrolyzer[29] under controlled conditions to give useful information. These data are somewhat similar to those obtained from the mass spectrometer.[4] All these advanced techniques are useful in identifying the unknown compound and should not be overlooked.[20]

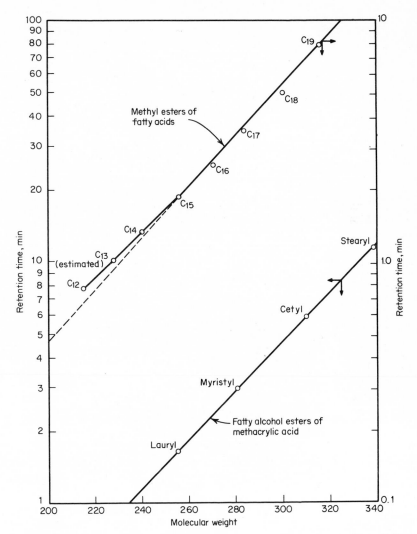

Fig. 4-50 Separation of fatty acid methyl esters and methacrylic acid esters of fatty alcohols on SE-30 and XE-60 column (2.5/0.5%) (6 ft × 3/16 in.) at 185°C vs. molecular weight.[16]

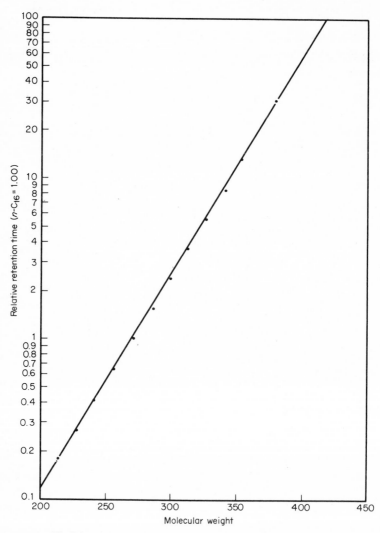

Fig. 4-51   *n*-Alkyl fatty acid methyl esters: molecular weight vs. relative retention time (n-C$_{16}$ = 1.00) on Apiezon M at 197°C.[7]

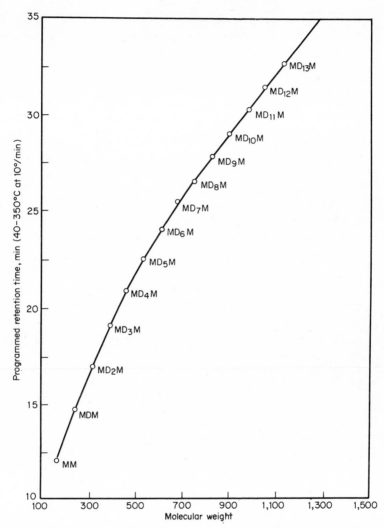

Fig. 4-52   Linear siloxanes: variation of molecular weight with programmed retention time on an SE-30 column (20%) (6 ft × ⅛ in.) from 40 to 350°C at 10°/min.[8]

REFERENCES

1. Alber, A.: *Ind. Eng. Chem., Anal. Ed.*, **12**: 764 (1940).
2. American Chemical Society: *Physical Properties of Chemical Compounds II, Adv. Chem. Ser.*, no. 22, 1959.
3. Beck, W.: *Facts Methods*, **8** (6): 9 (1967).
4. Biemann, K.: *Techniques of Organic Chemistry*, vol. 11, pt. I, pp. 260–316, Interscience Publishers, Inc., New York, 1963.
5. Bordenca, C.: *Ind. Eng. Chem., Anal. Ed.*, **18**: 99 (1946).
6. Burchfield, H. P., and E. E. Storrs: *Biochemical Applications of Gas Chromatogaphy*, p. 270, Academic Press Inc., New York, 1962.
7. *Ibid.*, p. 552.
8. Carmichael, J. B., and J. Heffel: *J. Phys. Chem.*, **69**: 2213 (1965).
9. Chamot, E. M., and C. W. Mason: *Handbook of Chemical Microscopy*, 3d ed., vol. I, p. 402, John Wiley & Sons, Inc., New York, 1958.
10. Dimick, K. P.: *Gas Chromatographic Preparative Separations*, pp. 8-6, 8-7, and 14-6, Varian Aerograph, Palo Alto, Calif., 1966.
11. Fajans, K.: *Physical Methods of Organic Chemistry*, 2d ed., vol. 1, pt. II, p. 1162, Interscience Publishers, Inc., New York, 1949.
12. Feibush, B., and E. Gil-Av: *J. Gas Chromatogr.*, **5**: 257 (1967).
13. Germann, F. E. E., and O. S. Knight: *Ind. Eng. Chem.*, **26**: 467 (1934).
14. Golay, M. J. E.: U.S. Pat. 2,920,478 (1960) (to Perkin-Elmer Corp.).
15. Gil-Av, E., B. Feibush, and R. Charles-Sigler: *Tetrahedron Lett.*, **10**: 1009 (1965).
16. Gudzinowicz, B. E.: *Application of Gas Chromatography to Pharmacy, Medicine, and Toxicological Analysis*, pp. 44–45, Jarrell-Ash Co., Waltham, Mass., 1964.
17. Hause, J. A., J. A. Hubicki, and G. G. Hazen: *Anal. Chem.*, **34**: 1567 (1962).
18. Kaufmann, H. C.: *Handbook of Organometallic Compounds*, p. 652, D. Van Nostrand, Company, Inc., New York, 1961.
19. Klages, F., and K. Mohler: *Ber.*, **81**: 411 (1948); **84**: 56 (1951).
20. Kobayaski, R., P. S. Chappelear, and H. A. Deans: *Ind. Eng. Chem.*, **59**: 63–82 (1967).
21. Littlewood, A. C.: *Gas Chromatography*, pp. 458–459, Academic Press Inc., New York, 1962.
22. McReynolds, W. O.: *Gas Chromatographic Retention Data*, p. 144, Preston Technical Abstracts Company, Evanston, Ill., 1966.
23. *Ibid.*, p. 260.
24. *Ibid.*, p. 232.
25. *Ibid.*, pp. 62, 198.
26. Meldrum, W. B., L. P. Saxer, and T. O. Jones: *J. Am. Chem. Soc.*, **65**, 2023 (1943).
27. Mengies, A. W. C., and S. L. Wright: *J. Am. Chem. Soc.*, **43**: 2314 (1921).
28. Nichols, J. B.: *Natl. Paint Bull.*, **1**: 12–14 (1937).
29. Perry, S. G.: *Adv. Chromatogr.*, **7**: 221–241 (1968); F. J. Kabot and N. B. Coupe: *Am. Lab.*, pp. 31–37 (October 1968).
30. Schneider, F. L.: *Qualitative Organic Micro-analysis*, pp. 69–71, Academic Press Inc., New York, 1964.
31. Skau, E. L., and H. Wakeham: in A. Weissberger (ed.), *Physical Methods of Organic Chemistry*, 2d ed., vol. 1, pt. I, chap. 3, Interscience Publishers, New York, 1949.
32. Sucharda, E., and B. Bobranski: *Chem. Z.*, **51**: 568 (1927).
33. Swietoslawski, W., and W. Romer: *Chem. Zentralbe.*, **1**: 2125 (1926).

34. Vogel, A. I.: *J. Am. Chem. Soc.*, **70**: 607–610 (1948).
35. Weast, R. C.: *Handbook of Chemistry and Physics*, 45th ed., The Chemical Rubber Publishing Co., Cleveland, Ohio, 1964.
36. *Ibid.*, table C.
37. Weatherall, L.: *J. Chromatogr.*, **27**: 35 (1967). (See also Wilkens Instrument Co.: *Aerograph Res. Notes*, Summer, 1965.)
38. Weissberg, S. G., R. Rothman, and M. Wales: in *Analytical Chemistry of Polymers*, vol. 12, G. M. Kline (ed.), pp. 29–46, Interscience Publishers, Inc., New York, 1962.
39. Weissberger, A.: *Practical Methods of Organic Chemistry*, vol. I, pp. 77–86, Interscience Publishers, Inc., New York, 1945.
40. Wilkens Instrument Co.: *Aerograph Res. Notes*, Fall, 1962, p. 2, Varian Aerograph, Palo Alto, Calif.

CHAPTER FIVE

Functional-Group and Classification Tests

LOGICAL ORDER OF PROCEDURE

A record should be made of all the information learned to date concerning the unknown compound, including all physical meaurements, gross observations, chemical properties, elemental composition, solubility properties, and gas-chromatographic behavior on various columns. Logically the next procedure is to determine what functional groups are actually present. Some clues from the behavior on various polar and nonpolar chromatographic columns should give a good indication of possible groups in addition to those found from the solubility studies.

To decide on the class or classes of compounds to which the unknown belongs, determine the exact nature of the functional groups present. The order in which the tests are made should be carefully planned so that the minimum number of tests will be needed.

1. A test that eliminates a class is just as important as one that reveals a class to be present.

2. A class of compounds commonly utilized in industry or in the laboratory is more often encountered than a less common class.

3.   The pH can be determined on a substance even though it may be only slightly soluble in water, but if the unknown is impure, the pH can be misleading, as traces of free acids may indicate a misleading acidic character.

4. Most carbon compounds are colorless.   Color is a good indication that certain functional groups are likely to be present.[149]   There are a few colored substances which contain carbon, hydrogen, oxygen, halogens, or sulfur, such as quinones, aromatic ketones with unsaturated side chains, and a few alkanediones.

Nitrogenous compounds, with or without halogens or sulfur, substituted anilines, toluidines, polycyclic amines or hydrazines, nitro-, nitroso-, or aminophenols, azo or diazo compounds, picrates, hydrazones, and osazones are colored.   The azo compounds are generally red or orange.   Yellow chromophoric groups are usually produced by $-N=N-$, $=C=S$, $=C=O$, $-N=O$, $-NO_2$, and the o- or p-quinone structures.   Color-intensifying groups (auxochromic) are usually $-NH_2$, $-NHR$, $-OH$, or $-SH$.   Halogenated nitro hydrocarbons display color, which is deepened as chlorine is replaced by bromine or iodine.

Unsubstituted anilines show no color unless they are exposed to air, when they tend to darken.   Purification removes the color.   Numerous other compounds develop color in air.

5. On the basis of the elements present, one should first investigate the *most probable combinations*.   For example, if the unknown contains carbon, hydrogen, nitrogen, and oxygen, it may be an amino acid; or if it also contains sulfur, it may be a sulfonamide.

By careful selections based on specific class or functional-group tests plus the classifications based on solubility and the elements present, one can logically determine what further tests are needed to verify the composition.

This section of the book is arranged according to elements present: the tests suggested for each group are found below each compound type in the section on functional tests.

Compounds Containing Only Carbon and Hydrogen

These compounds are generally insoluble or slightly soluble in water or most of the solubility-classification reagents.   Unsaturated or aromatic hydrocarbon compounds may react with cold concentrated or fuming sulfuric acid.

Most hydrocarbons are readily separable on the gas chromatograph. Classification reagents serve to assist in the separation of mixed unsaturated or aromatic hydrocarbons. For example, bromine in carbon tetrachloride reacts with unsaturates, producing brominated components which elute at higher temperatures (under programmed operation). The reactive unsaturated hydrocarbons disappear from the chromatogram but may appear at higher temperatures when programmed. If the reaction has taken place by substitution rather than addition, as with benzene or toluene, hydrogen bromide is evolved. This can be detected chromatographically as shown in Fig. 5-1, as well as by the fuming and decolorization of the bromine.

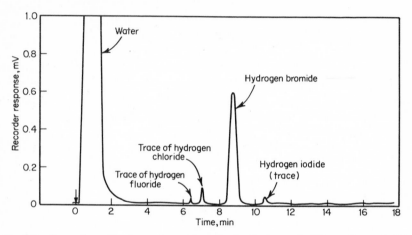

Fig. 5-1 Separation of hydrogen bromide and other hydrogen halides on Poropak QS (1½ ft × ¼ in.) at 80°C using gold-plated thermal-conductivity detector filaments.

Bromine water similarly will brominate by substitution and by addition, but the evolution of hydrogen bromide can be detected only by chromatographic instruments or by identification of the bromide ion in the water layer.

Aqueous potassium permanganate is decolorized by acetylenic or ethylenic groups, producing hydroxyl or higher oxidized groups (see glycols or ketones) at the unsaturated linkage. Further heating will oxidize and cleave the chain (see carboxylic acids). These fractions can be examined on the chromatograph, verification which can assist in establishing the location of the double bond. Sometimes certain activated side chains are easily oxidized by permanganate, e.g., oxidation of toluene to benzoic acid or methyl pyridines to acids. Insoluble unsaturated compounds will decolorize permanganate solution only if they are ground to a fine powder and shaken vigorously with the reagent.

Sodium metal reacts with hydrocarbons which have active hydrogens, producing a sodium compound of the hydrocarbon and free hydrogen. The hydrogen can be detected by injection of the trapped gas into a chromatograph (see hydrogen under reducing agents). The sodium compounds can be converted into volatile compounds by treatment with dilute acids or reaction with trimethylsilylating reagents.

In chromatographing mixed hydrocarbons on ethylene glycol columns treated with silver ion, the unsaturated hydrocarbons are adsorbed on the column, as shown in Fig. 5-2b compared with a conventional column without silver ions (Fig. 5-2a). Acetylenic hydrocarbons are strongly adsorbed, whereas ethylenic hydrocarbons are less strongly adsorbed.

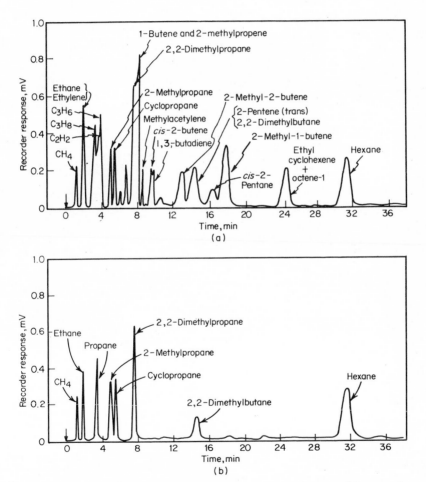

Fig. 5-2 Removal of saturated hydrocarbons by gas chromatographing on silver nitrate–treated polyethylene glycol columns, (a) before and (b) after.

Unsaturated hydrocarbons can be hydrogenated by injection over a precolumn of activated nickel or platinum in a stream of hydrogen. The unsaturated peaks disappear, and the saturated hydrocarbon peaks increase in size (see Fig. 5-3).

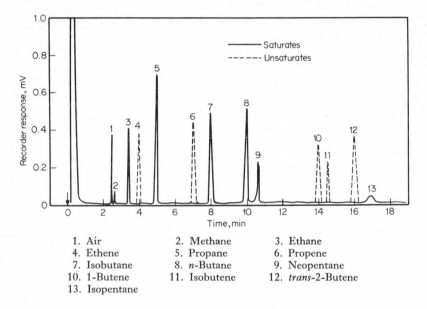

1. Air	2. Methane	3. Ethane
4. Ethene	5. Propane	6. Propene
7. Isobutane	8. n-Butane	9. Neopentane
10. 1-Butene	11. Isobutene	12. trans-2-Butene
13. Isopentane		

Fig. 5-3 Conversion of unsaturates to saturated hydrocarbons by hydrogenation on a nickel column, separated on propylene carbonate (20%) column (915 cm × 0.64 cm) at 26°C.

Straight-chain hydrocarbons produce straight lines when plotted against the carbon number. The branched hydrocarbons show similar straight lines with shorter retention times. There is less and less retention time with increasing branching. Aromatics and cyclic saturates produce straight lines for their own homologous series that are proportional to their boiling points (see Fig. 5-4).

The specific retention volumes or retention times are proportional to other physical properties, plotted for various homologous series of compounds in Chap. 4. Since refractive index is such an accurate property, more accurate estimates of where a given hydrocarbon should elute can be predicted from its refractive index. Similarly, the refractive index of the unknown can be obtained by plotting these values and extrapolating to the unknown in question.

Certain thermodynamic data have also been plotted versus the retention data. For uniformly increasing or decreasing data, the thermodynamic properties can be extrapolated to components that are

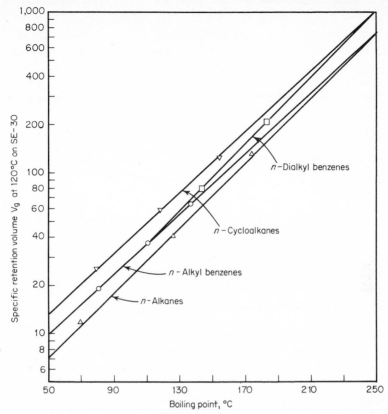

Fig. 5-4 Various hydrocarbons: relationship of boiling point to specific retention volume on an SE-30 column at 120°C.[125]

difficult to determine or very rare species (see Chap. 4).   For example, on a series of high-boiling homologous linear siloxanes, physical thermodynamic and chemical properties have been measured.   If some of these values were less accurate than others, plots of the physical or thermodynamic properties versus their programmed retention time would indicate erroneous values.

Similarly, physical and thermodynamic values can be predicted from previously measured properties of lower homologs.   Naturally, these values are only as accurate as the measured values of the lower homologs. Thus, the value of gas chromatography is extended by studying these physical, chemical, and thermodynamic values.   Its effectiveness is further enhanced by predicting data in areas beyond the range of present instrumentation or for isomers or homologs that can be separated only with difficulty.   Physical and chemical properties of unknown compounds may be estimated before enough of the substance has been isolated or synthesized to be tested directly.   For a thorough theoretical discussion the reader is referred to the text by Leibnitz and Struppe.[117]

In the solubility-classification tests, most hydrocarbons fall into one of two groups: soluble in concentrated sulfuric acid or insoluble in this reagent.

Hydrocarbons are partially identified by elimination.   Most hydrocarbons can be chromatographed on a variety of columns.[35,41,107,126,146,185] Apiezon L and SE-30 are recommended for n-alkanes.

ALIPHATIC HYDROCARBONS

Saturated

Aliphatic hydrocarbons are the most inert and least reactive of the hydrocarbons.   They do not give any positive test for any functional groups.   Most identification is obtained by physical-property measurements plus chromatographic examination.   Many commercial samples of hydrocarbons are mixtures.   The retention data of these mixtures can be plotted against their homologs according to boiling point (Fig. 5-4), carbon number (Fig. 5-5), refractive index (Fig. 5-6), density data (Fig. 5-7), or other physical and thermodynamic data (see Chap. 3).

Branching tends to reduce boiling points.   For example, a branched aliphatic hydrocarbon tends to elute before its n-alkyl homolog.   Infrared data, plus nmr spectra, greatly assist the analyst in verifying the identity of the saturated hydrocarbon, but chromatographic data plus physical measurements are the most convincing.

Unsaturated

The aliphatic hydrocarbons are usually soluble in the concentrated sulfuric acid solubility reagent.   They decolorize bromine in carbon tetrachloride or glacial acetic acid without evolution of hydrogen bromide, and they decolorize bromine water.   Potassium permanganate solution is usually rapidly decolorized (Baeyer's test).   It is quite possible for an unsaturated linkage to be sterically hindered so completely by branched groups that it fails to respond to these tests.   After the reaction with bromine or potassium permanganate is complete, the remaining residue is washed or the carbon tetrachloride removed by evaporation.   The resulting residue is examined by chromatography or other physical measurements.   It is possible to split the hydrocarbon at its point of unsaturation by vigorous oxidation.   The two fragments can be separated and examined further to determine the exact point of unsaturation.

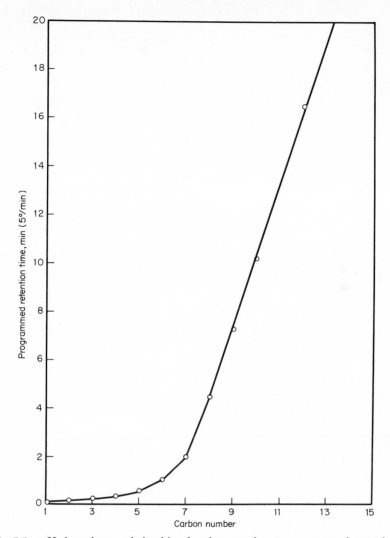

Fig. 5-5   *n*-Hydrocarbons: relationship of carbon number to programmed retention time on SE-30 column from 50 to 250°C at 5°/min.[40]

For example, an unsaturated hydrocarbon examined by the usual tests was found to contain one unsaturated bond.   Chemical analysis revealed that the compound had six carbons in its chain.   Gas chromatography showed that the compound eluted not before but after the *n*-alkyl saturate, indicating that the compound probably was not branched.   Nmr examination proved that it had only one terminal

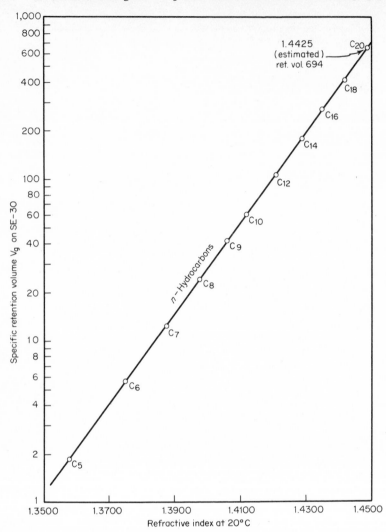

Fig. 5-6   *n*-Hydrocarbons: variation of refractive index at 20°C vs. specific retention volume on SE-30 column at 160°C.[116, 125]

methyl group.   From this information the probable composition could be estimated as follows:

$$CH_2= CH-CH_2CH_2CH_2CH_3$$

Oxidation produced two acids, one short-chain and one long-chain:

$$CH_2=CH-CH_2CH_2CH_2CH_3 \xrightarrow{\text{(O)}} HCOOH + HOOC(CH_2)_3CH_3$$

which on further examination can be shown to be formic and *n*-pentanoic acids.

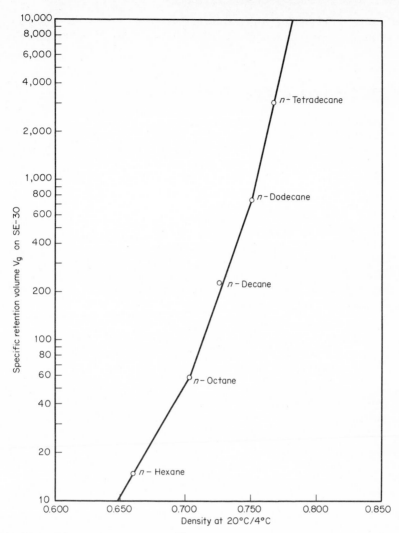

Fig. 5-7 n-Alkanes: relationship of d_4^{20} to specific retention volume on SE-30 column at 120°C.[116, 124]

If this compound is hydrogenated, a compound with a retention time identical to n-hexane is obtained.   Bromine produces a 1,2-dibromo-hexane:

$$CH_2{=}CH{-}CH_2CH_2CH_2CH_3 \xrightarrow{\ Br_2\ } CH_2BrCHBrCH_2CH_2CH_2CH_3$$

Many other tests can be performed on this compound, but usually the investigator has sufficient evidence of the probable identity of the

substance.   Several carefully chosen derivatives plus their chromatographic data and the physical measurements usually can accomplish the
final identification.

ALIPHATIC CYCLIC HYDROCARBONS

Saturated

Most of these hydrocarbons are higher-boiling than the corresponding
n-alkanes and have higher densities and refractive indices.   Their
retention data on such inert columns as SE-30 (silicone gum rubber) are
greater than those of their n-alkane homologs.   They do not give any
functional-group test and are difficult to nitrate or sulfonate.   They are
not easily oxidized or reduced except when the entire molecule is oxidized
or destroyed.   Investigators have oxidized some of these cycloparaffins
to alcohols and ketones under carefully controlled conditions.[65]

Since these hydrocarbons are insoluble in all the solubility-classification
reagents, they belong in the inert class of compounds.   Identification
can be made by measurements of physical properties and comparisons
of retention data on various columns with known substances. Nmr data
will verify whether they contain terminal methyls depending on whether
they have branched side groups.   Infrared data should serve to verify
the cyclic, branched, or straight-chain structures.

Cyclic hydrocarbons should behave like any other homologous series.
Branching produces another homologous series parallel to the parent
cyclic substances.

In certain disubstituted cyclic materials, two isomerically similar
compounds with cis and trans structures are obtained.   These are
separable on special columns (see Fig. 5-8).

Unsaturated

As unsaturated groups are introduced into aliphatic cyclic structures,
the substances become more aromatic in character.   These unsaturated
cyclic compounds are soluble in the concentrated sulfuric acid solubility
reagent and decolorize bromine in carbon tetrachloride without evolution of hydrogen bromide and bromine in water; potassium permanganate is similarly decolorized.

Usually the introduction of an unsaturated bond into the lower
cycloparaffins has a greater reduction in boiling point than the same
introduction in the higher homologs.   For example, cyclopentane has a
boiling point of 49°, cyclopentene 46°, cyclohexane 81°, and cyclo-

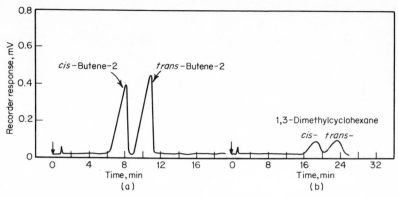

Fig. 5-8 Separation of cis and trans isomers at 40°C on: (*a*) 12 ft × ¼ in. benzyl ether column; (*b*) 8 ft × ¼ in. hexadecane column.

hexene 84°.   Introduction of the second unsaturated group has a greater effect on the boiling point; that is, the boiling point of 1,3-cyclo-pentadiene is 41°, and that of 1,3-cyclohexadiene is 80°.

Vigorous oxidation at the double bond will open the cyclic ring to form a dicarboxylic acid.   The dibromide formed by the addition of bromine to the double bond can be examined by gas chromatography (see Fig. 5-9).

These compounds are identified by their physical and chemical properties together with the chromatographic data.   Infrared, ultraviolet, and nmr data assist in making the final verification.

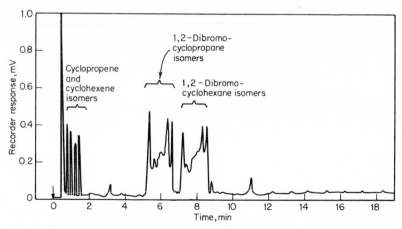

Fig. 5-9 Gas-chromatographic examination of a cyclohexane dibromide on a W-98 (10%) column (6 ft × ¼ in.) at 65 to 245°C.

AROMATIC HYDROCARBONS

With the introduction of a third double bond into 1,3-cyclohexadiene, the compound becomes distinctly aromatic in character:

Cyclohexane   Cyclohexene   1,3-Cyclohexadiene   1,3,5-Cyclohexatriene

                                                                        (benzene)

Increasing aromatic character $\longrightarrow$

The unsubstituted aromatic hydrocarbons are insoluble in cold concentrated sulfuric acid, but on warming the benzene will react. With the introduction of a methyl group, the ring is sulfonated easily at room temperature and the hydrocarbon becomes soluble in cold concentrated sulfuric acid.

In using classification and functional-group reagent tests, the aromatic hydrocarbons can be nitrated rather easily; bromine in carbon tetrachloride is decolorized, with evolution of hydrogen bromide; benzene does not decolorize potassium permanganate in the cold, but a toluene side chain is oxidized to benzoic acid. Almost all aromatic hydrocarbons can be chromatographed as shown in Fig. 5-10.

Even their reaction products can be chromatographed. Since cold concentrated sulfuric acid dissolves most aromatic hydrocarbons except

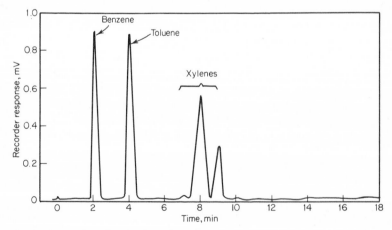

Fig. 5-10   Separation of aromatic hydrocarbons on W-98 (silic one gum rubber column, 10%) (6 ft × ⅛ in.) at 75°C.

benzene, shaking the reagent with the mixed hydrocarbon solvent will serve to remove these hydrocarbons and other unsaturates, leaving only aliphatic or saturated cycloparaffins.   Another treatment with bromine in carbon tetrachloride will remove benzene from its regular place in the chromatogram.   Typical chromatograms of mixed hydrocarbons are shown in Fig. 5-11 with the treatments used to differentiate the various components.

Identification here is also made with physical measurements together with the chromatographic data and reaction derivatives.

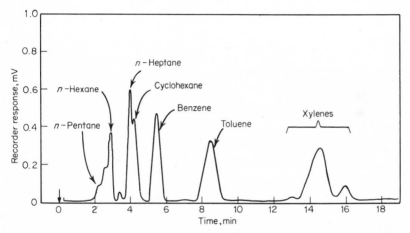

Fig. 5-11   Typical chromatogram of mixed hydrocarbons on a W-98 (10%) column (6 ft × ⅛ in.) at 75°C (sulfonation removes aromatics).

Compounds Containing Carbon and Oxygen or Carbon, Oxygen, and Hydrogen

Oxalic acid anhydride $(OOC-COO)_2O$ might be considered the first member of the carbon and oxygen series, but it is unstable except under special conditions.   There are a few compounds composed of only oxygen and carbon, such as CO, CO_2, and C_2O_3.   Most carbon and oxygen compounds which the analyst will be asked to identify have some hydrogen.

Compounds in this classification are usually alcohols, ethers, aldehydes, carboxylic acids, esters, anhydrides, ketones, carbohydrates, phenols, lactones, acetals, etc.   *Carboxylic acids* can usually be detected by their acidic reaction and tendency to be soluble to slightly soluble in water *and* soluble in alkali or sodium bicarbonate solutions.   Fatty acids can be separated on a variety of chromatographic columns (see Fig. 5-12).   Acid anhydrides separate as in Fig. 5-13.   All acids can be converted into esters and identified by the *hydroxamate test* or by chromatographing on polyester columns (see esters).   The acids can be converted into acid halides by use of phosphorus pentachloride or

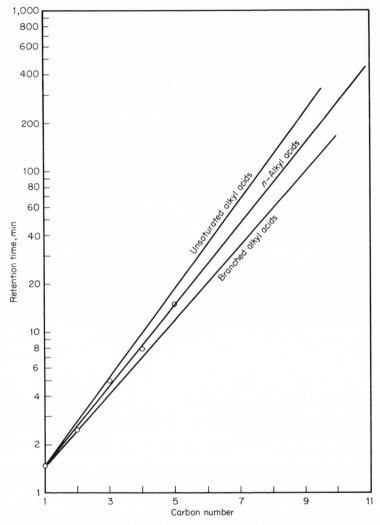

Fig. 5-12 Behavior of free carboxylic acids on Polypak 2 column (120/200 mesh) at 200°C (4 ft × ¼ in.).[92]

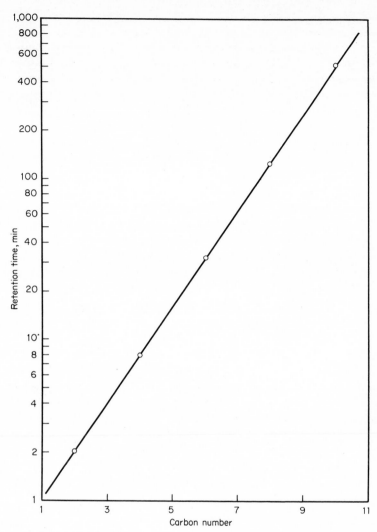

Fig. 5-13 Separation of acid anhydrides on Polypak 2 column (100/120 mesh) at 200°C (4 ft × ¼ in.).[92]

thionyl chloride (see halogen compounds) and chromatographed.   These acids may also be reacted with an alcohol to form esters on which the hydroxamate test may be performed and which are chromatographed:

$$C_4H_9COOH + SOCl_2 \longrightarrow C_4H_9COCl + SO_2 + HCl$$

$$C_4H_9COCl + C_5H_{12}OH \longrightarrow C_4H_9COOC_5H_{12} + HCl$$

ANHYDRIDES

Many anhydrides chromatograph directly on nonreactive columns (such as SE-30, silicone gum rubber, or Poropak) (see Fig. 5-13) and may be converted directly to hydroxamic acids without ester formation:

$$(CH_3CH_2CO)_2O + H_2NOH \longrightarrow C_2H_5COOH + C_2H_5NHOH$$

For this purpose, add a drop or two of the anhydride to $\frac{1}{2}$ ml of 1 N hydroxylamine in methanol and heat to boiling.   Cool and add a drop of 10% ferric chloride solution.   If hydroxamic acid is present, a wine-red colored complex will develop.

If an anhydride is suspected, even though the above test gave a negative response, heat a few drops of the anhydride with a similar quantity of n-butanol.   Cool and apply the hydroxamate test.   Sometimes, due to the slowness of the reaction with hydroxylamine (steric hindrance, inhibition, etc.), the test may appear negative until the ester is formed.

Similarly, many acyl or aroyl halides will give definite positive reaction with the anhydride test, but one should be able to suspect the presence of halogens.   To verify, run the halides test as described, which should give different performances on the various chromatographic columns.

ALDEHYDES AND KETONES

Unfortunately, many aldehydes and ketones give similar tests based on reaction with the carbonyl group.   The tests that are specific for aldehydes but not for ketones are described below under the classification tests for aldehydes (see Schiff's, benzidine, methone, and condensation tests).   Aldehydes chromatograph easily, as shown in Fig. 5-14.

Phenylhydrazine hydrochloride reacts with either aldehydes or ketones. Similarly, 2,4-dinitrophenylhydrazine reacts with aldehydes or ketones.

Benzene sulfonhydroxamic acid is more specific for aldehydes, but o-nitrobenzaldehyde and p-hydroxy aryl aldehydes do not respond to this test.   Benzyl ketones do give the positive reaction.   The reaction is suggested as follows (the positive test is the wine-red color with ferric chloride):

$$C_6H_5SO_2NHOH + 2KOH \longrightarrow C_6H_5SO_2K + KNO + 2H_2O$$

$$KNO + CH_3CHO + HCl \longrightarrow CH_3CONHOH + KCl$$

KNO is the salt of the hypothetical nitrosyl acid, HN=O, neither of which has been isolated to date.   To perform the test, dissolve 1 to 2 mg of benzene sulfonhydroxamic acid in $\frac{1}{2}$ ml of methanol and add 10 mg of the aldehyde and $\frac{1}{2}$ ml of 2 N potassium hydroxide in methanol.

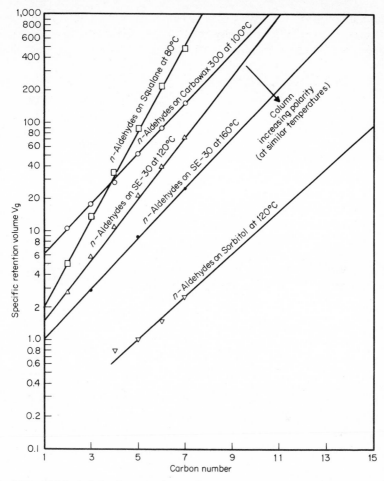

Fig. 5-14 Aldehyde behavior on various polar and nonpolar columns at 80, 100, 120, and 160°C.[125]

Boil, cool, and add 1 drop of 10% $FeCl_3$ solution. The wine-red color will result if aldehydes are present.

Schiff's test is positive for aldehydes. *p*-Rosaniline hydrochloride is reduced with sulfurous acid to produce a leucosulfonic acid of the dye. The reduced dye reacts with more sulfurous acid to form the colorless bis (*N*-sulfinic) acid; 2 mol of the aldehyde react with the acid dye, causing it to lose 1 mol of sulfurous acid and producing a reddish-purple quinoid dye complex. Some ketones give a faint color with this reaction. A few compounds other than aldehydes give light pink colors; these color reactions are not characteristic of aldehydes unless they have a bluish cast. Most of these derivatives are too unstable to be chromatographed but may be paper-chromatographed.   To perform the test add 3 drops of the aldehyde to be examined to 2 ml of Schiff's reagent.   Without warming, a purplish red color will develop within 10 min.

The *benzidine test* will give a yellow, orange, or red color or precipitate with almost all aldehydes when reacted with benzidine in glacial acetic acid.   Alkanals tend to form yellow colors which change to red on heating; aromatic aldehydes tend to form crystalline products with this reagent.   Use 3 ml of the freshly prepared reagent (3 to 4%) in glacial acetic acid with 3 drops of the aldehyde to be tested.   Some of these derivatives can be·chromatographed (see Fig. 5-15).

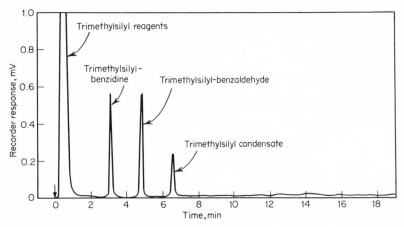

Fig. 5-15   Separation of benzidine products of aldehydes on W-98 column (10%) (6 ft × ¼ in.) programmed from 50 to 250°C at 10°/min.

The *methone test* does not give a positive test for ketones, only with aldehydes.[188]   A milky precipitate or suspension forms instantaneously when 50 mg of the aldehyde in 1 ml of water is mixed with 3 drops of 5% methone in ethanol.   Shake vigorously; the opalescent suspension should form within 2 min to give a positive test for aldehydes.

The suspension gradually crystallizes and forms colorless crystals. Some of the lower aldehydes form derivatives that can be chromatographed (see Fig. 5-16).   The higher aldehydes can be made to move more readily on the chromatograph by further conversion to trimethylsilyl derivatives.

Oximes, condensate products of hydroxylamine with an aldehyde or ketone, produce some derivatives which give sharp melting points or which can be chromatographed as is or as trimethylsilyl derivatives (see Fig. 5-17).

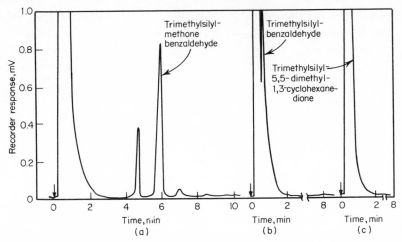

Fig. 5-16   Separation of some methone derivatives on W-98 column (10%) (6 ft × ¼ in.) at 250°C.

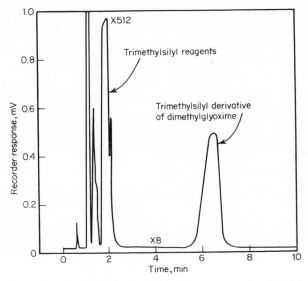

Fig. 5-17   Separation of oximes of aldehydes and ketones as trimethylsilyl derivatives on SE-30 column (20%) (5 ft × ⅛ in.) programmed from 100 to 250°C at 10°/min.

Oxidation of Aldehydes

Aldehydes can be easily oxidized to their corresponding carboxylic acids. The aliphatic aldehydes are more easily oxidized than the corresponding aromatic aldehydes. Only a few ketones are oxidized by such mild oxidizing agents as cupric ion or silver ion in an alkaline solution (see Benedict's, Fehling's, and Tollens' tests).

Examine the products of the oxidation by the gas chromatograph. These acids may be examined after conversion to suitable esters.

Benzaldehyde, in the presence of alkali, automatically oxidizes one part to benzoic acid and reduces another part to benzyl alcohol (Cannizzaro reaction) (see Fig. 5-18).

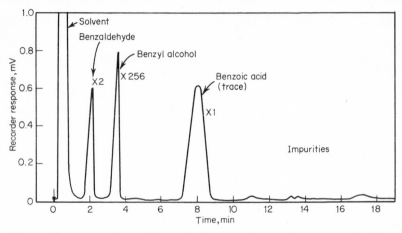

Fig. 5-18 Separation of benzaldehyde and its reduction product, benzyl alcohol, on polyphenyl ether (PPE-20) column (3%) (6 ft × ⅛ in.) at 125°C.

Reduction of Aldehydes

Aldehydes can be reduced to alcohols by hydrogen catalyzed with finely divided nickel.   This may be done in a hydrogenation bomb or by injecting the aldehyde into a precolumn of activated nickel using a stream of hydrogen.   As shown in Fig. 5-19, the aldehyde disappears, and the corresponding alcohol takes its place.

For aldehydes that are not too sensitive to acids, the compound may be reduced with zinc and hydrochloric acid, iron and hydrochloric acid, or tin and hydrochloric acid (see Fig. 5-20).

Condensation of Aldehydes

When a solution of an aldehyde, such as acetaldehyde, is treated with sodium hydroxide or other alkaline catalyst, the two molecules condense, producing an aldol:

$$CH_3CHO + H\!-\!CH_2CHO \xrightarrow{\ OH^-\ } CH_3CHCH_2CHO$$

$$\underset{\text{Aldol}}{\overset{|}{OH}}$$

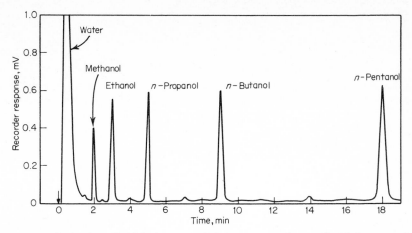

Fig. 5-19   Reduction of aldehydes to alcohols over a precolumn of nickel separated on a Poropak Q column (6 ft × ¼ in.) at 150°C.

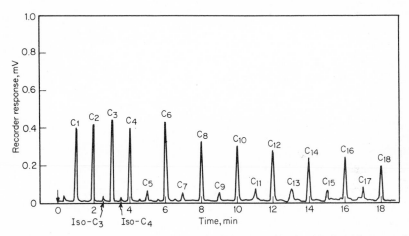

Fig. 5-20   Reduction of aldehydes to alcohols with zinc and hydrochloric acid separated on Poropak Q column (6 ft × ¼ in.) at 150°C.

This reaction takes place only with aldehydes having at least one hydrogen in the position α to the carbonyl group of the adding molecule. Trimethylacetaldehyde, $(CH_3)_3CCHO$, and benzaldehyde, C_6H_5CHO, *do not* show aldol condensation. Propionaldehyde, CH_3CH_2CHO, gives the aldol condensation through its α hydrogens rather than a β hydrogen:

$$CH_3CH_2CHO + \underset{\underset{\displaystyle CH_3}{|}}{CH_2CHO} \xrightarrow{\ OH^-\ } CH_3CH_2CH\underset{\underset{\displaystyle OH}{|}}{-}CH\underset{\underset{\displaystyle CH_3}{|}}{-}CHO$$

These aldols can eliminate water rather easily under the influence of an acid catalyst:

$$\underset{\substack{| \quad | \\ OH \quad H \\ \text{Aldol}}}{CH_3CH-CHCHO} \xrightarrow{-H_2O} \underset{\text{Crotonaldehyde}}{CH_3CH=CHCHO}$$

Acetone will undergo aldol condensation under special conditions.[58]

ALCOHOLS

Alcohols are easily chromatographed on a variety of columns. On certain polyester columns, some high-molecular-weight alcohols will exchange, especially in the presence of acids or alkaline catalysts (see Fig. 5-21).

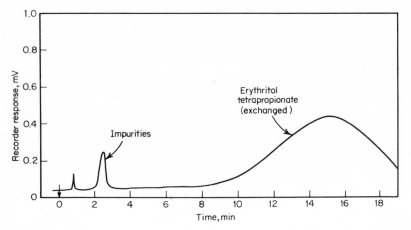

Fig. 5-21   Ester exchange on polyester column (sucrose octaacetate) with erythritol tetrapropionate at 150°C.

Esterification

Alcohols are readily esterified by reaction with acid chlorides or with the free acid in the presence of a catalyst (such as BF_3 or BCl_3). They may be ester exchanged by means of an acid or base catalyst.

Tertiary alcohols are only partially reacted to form alkyl chlorides when mixed with acetyl chloride, due to the liberation of hydrogen chloride. An acid acceptor, such as dimethylaniline or pyridine, may be added to the reaction mixture to accept the liberated hydrogen chloride.

Add 0.2 ml of the unknown tertiary alcohol to a mixture of 0.1 ml of acetyl chloride and 0.1 ml of dimethylaniline or pyridine. Shake the mixture vigorously for 5 min; add 5 g of ice to destroy excess acetyl chloride.   Test the oily layer for esters by means of the hydroxamate test and chromatograph for comparison with known esters.

Xanthate Test

Alcohols may be reacted as the potassium alkoxides with carbon disulfide to yield potassium alkyl xanthates as yellow precipitates.   The mono-alkyl ethers or glycols yield a satisfactory test, but the monoalkyl ethers of diethylene glycol produce red oils instead of the characteristic yellow insoluble products.   The tertiary alcohol xanthates hydrolyze easily but usually produce enough insoluble yellow precipitate to give a positive test.

To the 0.5 ml of the alcohol in a test tube, add a pellet of potassium hydroxide and heat until dissolved.   Cool and add 1 ml of ether followed by dropwise addition of carbon disulfide until 0.5 ml has been added.   If no yellow precipitate is produced after 0.5 ml of CS_2 solvent has been added, the material is not an alcohol.

To confirm the presence of xanthate, react 5 mg of the insoluble product with several drops of a solution containing 1 mg of ammonium molybdate.   A reddish-blue color develops upon addition of 4 drops of 2 N hydrochloric acid to form $(ROCOSH)_2MoO_3$ complex.

If the xanthate is dissolved in a suitable solvent and injected into the chromatograph, the unstable compound yields carbon disulfide plus the alcohol (see Fig. 5-22).

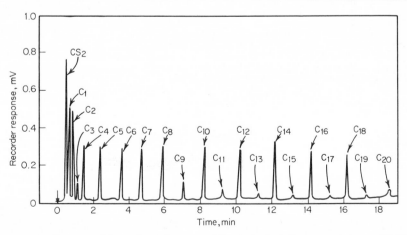

Fig. 5-22 Decomposition of xanthates to CS_2 and alcohols, separated on W-98 (10%) column (6 ft $\times$ ¼ in.) programmed from 80 to 240°C at 10°/min.

Differentiation of Primary, Secondary, and Tertiary Alcohols (Lucas' Reagent[120])

This reaction depends upon the rate of conversion of the various alcohols to alkyl chlorides when subjected to the Lucas' reagent test (zinc chloride anhydrous in concentrated hydrochloric acid, 160 g/100 ml). Since the test relies upon the separation of a second liquid phase of alkyl chloride in the reagent, it is valid only for alcohols which are soluble in the reagent. This limits its use to alcohols which have one functional group and an alkyl group of hexyl or less. There are a few polyfunctional alcohols that give positive results.

To perform the test, mix 0.5 ml of the unknown alcohol with 3 ml of Lucas' reagent in a stoppered tube. Shake vigorously and allow to stand in a water bath at 25 to 30°C. If a reaction takes place, a milky suspension is noted due to the formation of an insoluble alkyl chloride. Tertiary alcohols react within 1 to 2 min, with separation of an oily layer later on. The separated oily layer can be injected into the instrument and compared with known alkyl halides, as shown in Fig. 5-23.

Secondary alcohols are of intermediate reactivity between the tertiary and primary alcohols. Concentrated hydrochloric acid does not produce the alkyl chlorides in appreciable quantities alone; they are converted fairly rapidly (within 5 min) in the presence of the zinc chloride. Within 5 min, a cloudiness is noted, and a distinct oily layer separates later (within 10 min).

If there is any question whether the unknown is a secondary or tertiary alcohol, repeat the test with another 0.5-ml sample and 3 ml of concentrated hydrochloric acid. Secondary alcohols do not show any appreciable reaction, whereas tertiary alcohols are converted to alkyl chlorides within 10 min with the concentrated acid.

Primary alcohols usually require more than 10 min to convert to the alkyl chlorides under the influence of Lucas' reagent. If the cloudiness appears immediately but no distinct layer forms, the tertiary alcohol may have been an impurity in the less reactive alcohol.

The behavior of allyl alcohol is different from that of the primary alcohols because of a more stable group. This alcohol reacts rapidly with Lucas' reagent with production of heat. If the reagent is diluted with ice water, the allyl chloride separates out and can be chromatographed (see Fig. 5-24).

To speed up the reaction with Lucas' reagent and primary alcohols, the solution must be warmed or allowed to stand at higher temperature in a water bath. Use of hydrogen bromide or hydrogen iodide (especially in the presence of the appropriate zinc halide) produces alkyl bromides and iodides much more readily than the alkyl chlorides of the primary alcohols. These alkyl halides yield similar homologous series of curves, as illustrated in Fig 5-25.

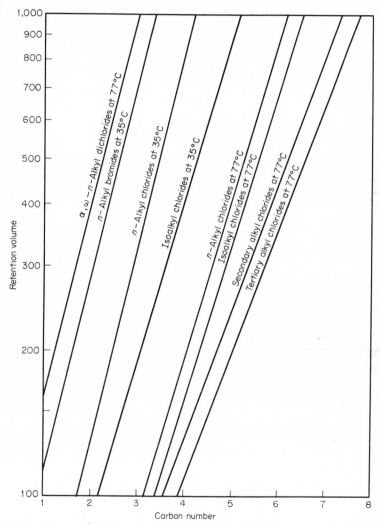

Fig. 5-23 Separation of alkyl halides on tricresyl phosphate (40%) column on Celite 545 (18 m × 4 mm) at 35 and 77°C.[88]

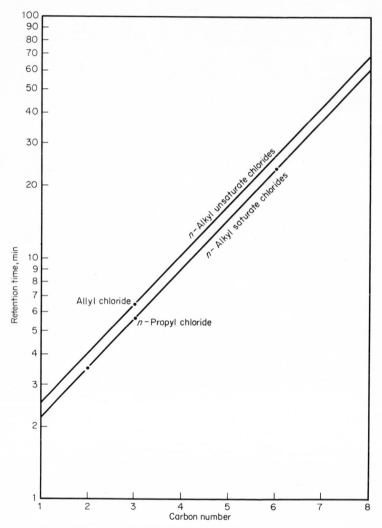

Fig. 5-24 Gas-chromatographic separation of allyl chloride and other chlorides on W-98 (10%) column (6 ft × ¼ in.) at 125°C.

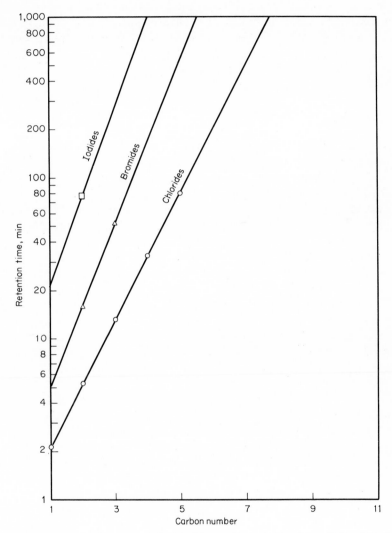

Fig. 5-25 Behavior of separated alkyl halides on Poropak Q column (6 ft × ⅛ in.) at 150°C.

CARBOHYDRATES

Because carbohydrates usually decompose easily in the injection port of the gas chromatograph, they must be converted into more stable derivatives in order to reduce their heat sensitivity. Certain groups are more sensitive to oxidation or reduction, which aids in differentiating certain closely related carbohydrates. Formation of acetates is useful for carbohydrates up to disaccharides; trimethylsilyl derivatives are used for higher polysaccharides.

Water-soluble

Monosaccharides are usually more easily oxidized than the disaccharides especially by the copper(II) ion in an acid solution

Barfoed's Test. Dissolve 16.6 g of copper(II) acetate crystals in 245 ml of distilled water. Then add 2.4 ml of glacial acetic acid. Mix 1 ml of this reagent with 1 ml of a 5% solution of the unknown carbohydrate in water. Immerse the tube with the mixture in a boiling water bath for 2 to 5 min. If the sugar is a monosaccharide, a reddish copper(I) oxide precipitate will form within 2 min. Any precipitate that forms after 2 min may be considered as being due to disaccharides or higher, since they reduce this reagent more slowly.

The solution from Barfoed's test may be evaporated to almost dryness on a steam bath and treated with acetic anhydride and pyridine (50/50). Allow the reagent to remain on the steam bath for 1 hr, cool, and inject into the instrument. Compare the results with those obtained by the same pyridine–acetic anhydride reagent before oxidation (see Fig. 5-26).

Benedict's Test. Dissolve 173 g of sodium citrate and 100 g of anhydrous sodium carbonate in 800 ml of warm distilled water. Pour a solution of 17.3 g of copper sulfate, $CuSO_4 \cdot 5H_2O$, in 100 ml of distilled water into the citrate-carbonate solution with continuous stirring. Cool and make up to 1 l. To 0.2 g of the unknown carbohydrate in 5 ml of distilled water, add 5 ml of Benedict's reagent. Heat to boiling; within 2 min a precipitate will form of copper(I) oxide, which may be red, yellow, or green, depending upon the amount of reducing carbohydrate present. Evaporate the oxidized carbohydrate to near dryness and treat with excess pyridine–acetic anhydride reagent to convert to the acetates for chromatographing. Compare with the carbohydrate acetates before oxidation (refer to Fig. 5-26).

Fehling's Test

Solution A. Dissolve 34.6 g of copper sulfate, $CuSO_4 \cdot 5H_2O$, in 500 ml of distilled water.

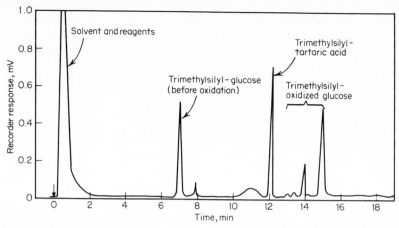

Fig. 5-26 Separation of trimethylsilyl derivatives of glucose before and after oxidation gas-chromatographed on XE-60 (10%) column at 150°C (6 ft × ⅛ in.) (acetates behave similarly).

Solution B. Dissolve 173 g of sodium potassium tartrate and 70 g of sodium hydroxide in 500 ml of distilled water.

To 0.2 g of the unknown carbohydrate in 5 ml of distilled water add 5 ml of a mixture (1/1) of the Fehling's solutions. Heat solution to boiling. A precipitate of copper(I) oxide will form if reducing groups are present in the carbohydrate. The neutralized solution may be filtered, evaporated to small volume on a steam bath, and treated with excess pyridine–acetic anhydride for chromatographic examination as described under Benedict's or Barfoed's test.

Differentiation of Aromatic and Aliphatic Aldehydes. By using both Benedict's and Tollens' tests, aromatic aldehydes can be distinguished from aliphatic aldehydes. Benedict's test is positive for aliphatic aldehydes, whereas Tollens' reagent is positive for all aldehydes.

Tollens' Reagent. Put 2 ml of silver nitrate solution (5%) in a clean test tube. Add 1 drop of 10% sodium hydroxide solution. Add dropwise ammonium hydroxide solution (2%) until the precipitate of silver oxide just dissolves on shaking. Prepare this solution immediately before use. Destroy all excess reagent, as explosive precipitates may form on standing.

Mix the reagent with the solution of the unknown carbohydrate. If a precipitate does not form immediately, warm slightly. Diphenylamine and other aromatic amines give a positive test, as do α-naphthol and certain other phenols. Leonard and Gelfand[119] have found that α-alkoxy and α-dialkylamino ketones reduce Tollens' reagent.

Similarly, the oxidized aldehydes may be examined by chromatography, as is or after converting the acids to methyl esters by use of methanol–boron trifluoride reagent (see Fig. 5-27). The alcohol groups may be acetylated with pyridine–acetic anhydride reagent (see under Barfoed's test).

Ketoses. Using Seliwanoff's test, ketoses are converted to hydroxymethylfurfural, which is condensed with resorcinol to form a colored

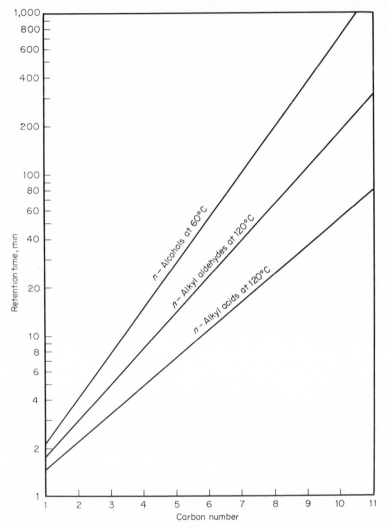

Fig. 5-27 Relationship of *n*-alkyl aldehydes and alcohols oxidized to acids separated on Reoplex 400 (10%) column treated with terephthalic acid (6 ft × ¼ in.) at 60 and 120°C.

organic complex.   To prepare the reagent, dissolve 0.125 g resorcinol in 250 ml of dilute hydrochloric acid (83 ml of concentrated acid mixed with 167 ml of distilled water).

To 1 ml of a 5% solution of the unknown carbohydrate in water, add 1 ml of Seliwanoff's reagent.   Heat to boiling; if the unknown is a ketose, a red color develops within 2 min.   Aldoses will develop a color on longer boiling or longer standing.

The hydroxymethylfurfural may be converted to an acetate or a trimethylsilyl derivative and chromatographed (see Fig. 5-28).

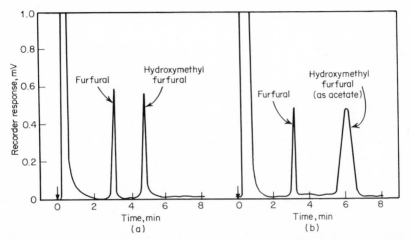

Fig. 5-28   Separation of hydroxymethylfurfural (*a*) as is and (*b*) as acetate derivatives on W-98 (10%) column (6 ft × ¼ in.) at 150°C.

Pentoses. Pentoses react in the presence of hydrochloric acid to form furfural, which in turn condenses with phloroglucinol to produce a red complex.   Other nonpentoses will produce yellow, orange, or brown complexes.

Dissolve approximately 10 mg of the carbohydrate under investigation in 5 ml of 6 N hydrochloric acid and add 10 mg of phloroglucinol.   Heat the mixture to boiling and hold at the boiling point for 1 min.   A red color indicates a pentose.

The boiled solution with hydrochloric acid alone may be examined by chromatography for furfural (see Fig. 5-29) or converted to derivatives and examined.

Water-insoluble

Iodine Test. Suspend 10 mg of the unknown carbohydrate in water and add several drops of saturated aqueous iodine solution.   Starches give a blue color; glycogen and higher dextrins produce reddish colors;

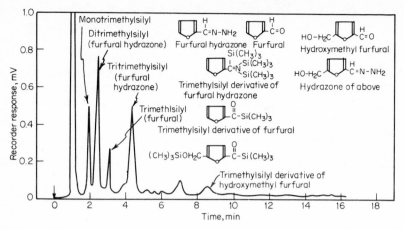

Fig. 5-29 Separation of furfural as trimethylsilyl derivatives on W-98 (10%) column (6 ft × ¼ in.) at 150°C.

inulin and the lower dextrins do not produce any colored complexes. Benedict's test is negative with all these substances until hydrolyzed. Most of the insoluble carbohydrates have too high molecular weights to produce volatile derivatives except trimethylsilyl derivatives. Hydrolysis of these materials usually produces substances volatile enough to prepare acetates or the trimethylsilyl derivatives. Some investigators have determined the type of starch present from microscopic examination.[31]

Hydrolysis Test. Suspend 100 mg of the substance in 15 ml of water and shake vigorously. Glycogen forms an opalescent solution before heating; starch forms an opalescent solution after heating. Add 0.5 ml of dilute hydrochloric acid and boil the mixture for at least 10 min. Cool, neutralize the solution, and test for reducing sugars, etc. The hydrolyzed substances can be evaporated to a small volume and treated with pyridine–acetic anhydride reagent or trimethylsilyl reagents to produce derivatives that can be examined by the chromatograph.

ESTERS

Most normal esters can be chromatographed as is on a variety of columns. However, polyesters, such as diethylene glycol succinate, cannot be chromatographed until they are degraded to the glycol and the diacid. The polyesters are usually too nonvolatile to be examined in the instrument. Becher and Birkmeier[9] have utilized polyesters as the nonvolatile liquid phase to coat the inert packing.   Injection of ethanol and hexane gives a relative measure of the polarity of the coating.

Hydrolysis

The esters are hydrolyzed with potassium hydroxide and diethylene glycol.   After hydrolysis is complete, the alcohol is distilled out of the solution and examined by the techniques used for hydroxyl compounds as well as by chromatography.   The alkaline portion is neutralized with 6 N sulfuric acid.   If volatile enough, this acid can be steam-distilled or extracted with chloroform and a derivative prepared.   The solution can be chromatographed for free acids on a nonpolar column such as SE-30 (see Fig. 5-12).

Hydroxamate Test

To 1 to 2 drops of the unknown ester add 0.5 ml of 1 N hydroxylamine hydrochloride in methanol.   Add 2 N potassium hydroxide in methanol to make solution alkaline to litmus paper.   Boil the mixture and cool. Acidify with dilute hydrochloric acid and add 1 drop of ferric chloride solution (10%).   A wine-red color indicates that an ester is present.

The hydroxamate ester may be chromatographed under the conditions shown in Fig. 5-30.

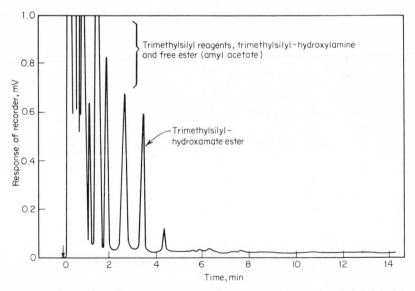

Fig. 5-30   Separation of hydroxamate ester of *n*-pentyl acetate as trimethylsilyl derivatives on W-98 (10%) column (6 ft × ⅛ in.) programmed from 100 to 250°C at 10°/min.

ETHERS

Almost all ethers can be chromatographed without alteration on a variety of columns. Only the high-molecular-weight (relatively non-volatile) ethers cannot be chromatographed without chemical alteration. An inert column, such as SE-30, can be used for separation of ethers; Carbowax may assist in separating polar compounds, e.g., ether and alcohols, from relatively nonpolar materials such as hydrocarbons (see Fig. 5-31).

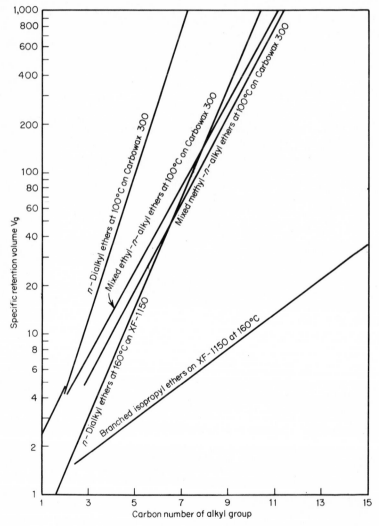

Fig. 5-31   Behavior of ethers on various columns (6 ft × ⅛ in.) at 100 and 160°C.

Esterification

A variety of ethers may be hydrolyzed and then transformed into acetate esters by reacting the ether with acetic acid and concentrated sulfuric acid. Gentle heating converts enough of the ether to an ester to give a positive hydroxamate test for esters.

To perform the test, to 0.5 ml of the unknown ether add 2 ml of glacial acetic acid and 0.5 ml of concentrated sulfuric acid. Reflux the mixture for 5 min and distill off several drops. Perform the hydroxamate test as described under esters. If the droplets do not give a positive test for esters (wine-red color), cool the refluxed mixture and add 5 ml of ice or ice water. Test the separated phase for esters. If no phase separates, shake out the liquid with ½ ml of benzene to extract the possible esters and recheck for esters by the hydroxamate test. Both the distillate and extracts may also be examined on the instrument, comparing the chromatogram obtained with the original ether before treatment.

Zeisel's Test

Hydriodic acid in glacial acetic acid reacts with alkyl ethers to give alkyl iodides. The acid will also react with alcohols, aryl alkyl ethers, esters, and acetals:

$$ROR' + 2HI \longrightarrow RI + R'I + H_2O$$
$$HOR' + HI \longrightarrow R'I + H_2O$$
$$ArOR' + HI \longrightarrow R'I + ArOH$$
$$RCOOR' + HI \longrightarrow R'I + RCOOH$$
$$RCH(OR')_2 + 2HI \longrightarrow 2R'I + RCHO + H_2O$$

To perform the test, place 0.1 g of the compound under test in a 16×150 mm test tube. Add 1 ml of glacial acetic acid and 1 ml of hydriodic acid (57%) plus a few boiling stones. Insert a gauze plug (described below) into the mouth of the tube (4 cm from mouth). Insert a small plug of cotton and tamp to 2 to 3 mm thick. On top of this plug place a piece of filter paper 2×10 cm folded lengthwise and moistened with mercuric nitrate solution (saturated: 1 ml of nitric acid plus 49 ml of distilled water). Immerse the tube 4 to 5 cm in an oil bath held at 120 to 130°C.

As the vapors rise through the plug, the alkyl iodides react with the mercuric nitrate, producing light orange to vermilion mercuric iodide. The color should develop within 10 min of heating. A yellow color is usually considered doubtful or negative.

To prepare the *gauze plug*, add 60 ml of 1 N sodium hydroxide solution to 1 g of lead acetate in 10 ml of distilled water with thorough stirring.   To this mixture, add 5 g of sodium thiosulfate crystals in 10 ml of distilled water.   Then add 1 ml of glycerol and dilute to 100 ml. Add 5 ml of this solution to strips of double cheesecloth 2×45 cm. Dry and roll to fit test tube.

Various investigators[1] have examined both the distillate and the pot residue by gas chromatography.   Mixed ethers were found and their alkyl iodides chromatographed (see Fig. 5-32).   Ethers containing more

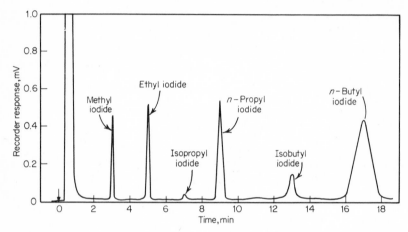

Fig. 5-32 Chromatogram of alkyl iodides from Zeisel degradation of mixed ethers (conditions similar to those of Fig. 5-25).

than four carbons were investigated by this technique, although much longer heating times are required.   Phenol (0.1 g), 2 ml of hydriodic acid, and 1 ml of propionic anhydride (in place of the glacial acetic acid) are added.   Upon degradation, glycerol yielded isopropyl iodide, as did propylene glycol; 1,3-propanediol yielded both isopropyl iodide and 1,3-propane diiodide; 1,4-butanediol yielded 1,4-butane diiodide. When polyethers of ethylene or propylene glycol were degraded by this reagent, methyl and ethyl iodide were obtained with the ethylene glycol polyether; methyl, ethyl, and isopropyl iodides were obtained with the propylene glycol polyether (see Fig. 5-33).

The above test, with chromatographic examination, would give negative results for ethers above C_4 alkyl groups due to the low volatility of the alkyl iodides.

The above procedure has been extended to alcohols, esters, mixed alkyl aryl ethers, and acetals.   Diaryl ethers are usually rather difficult to cleave by this technique, and more vigorous conditions are required.

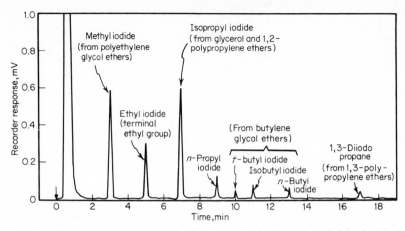

Fig. 5-33 Chromatogram of glycerol, glycols, and polyethers degraded by hydriodic acid (conditions similar to those of Fig. 5-32).

Iodine Test

To 0.5 ml of the ether add 1 ml of a light purple solution of iodine in carbon disulfide.   Ethers change the color of the solution to tan, aromatic hydrocarbons do not affect the purple color, and noncyclic hydrocarbons produce a light tan color without removing the purple color of the iodine solution.   Unsaturated hydrocarbons or unsaturated ethers that add iodine to their double bonds tend to destroy the iodine color.

PHENOLS

In the scheme of analysis, if the solubility tests indicate that a phenol could be present and the molybdic acid, ceric nitrate, ferric chloride, or nitrochromic acid tests are positive, one or more of the following may be used to confirm its presence.

Complex Indicator

Many phenols condense with phthalic anhydride to form indicators that produce red, green, blue, indigo, and violet colors in alkaline solutions. These solutions can change colors upon acidification.   Para-substituted phenols will not react to this test.

To 100 mg of the unknown phenol add 300 mg of phthalic anhydride and 200 mg of anhydrous zinc chloride in a test tube.   Heat mixture for 1 min to fuse the mass and cool.   Add a few milliliters of sodium hydroxide solution (1 to 2%), stopper, and shake.   Note the color when the solution is alkaline; acidify and note whether there is any change.

Many of these complexes are not volatile enough to be observed in the gas chromatograph unless further converted to more volatile substances (see Fig. 5-34).

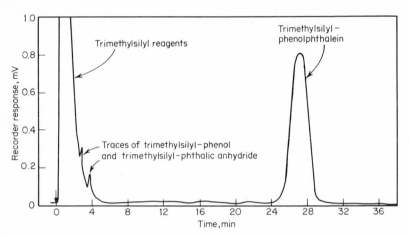

Fig. 5-34 Separation of trimethylsilyl phenolphthalein on W-98 (10%) column (6 ft × ⅛ in.) at 250°C.

Nitrous Acid Test

To 50 mg of the unknown phenol, add 1 ml of cold concentrated súlfuric acid and 10 mg of sodium nitrite.   Shake vigorously and warm gently.   A blue or purple color indicates a positive test for phenol. Slowly pour the mixture into 5 ml of distilled water.   The color will change to a purplish red.

The nitrous acid test depends upon the interaction of the phenol with nitrosophenol

$$C_6H_5OH + HONO \longrightarrow ONC_6H_4OH \rightleftharpoons HON{=}C_6H_4{=}O$$

thus producing an isomeric quinoid-type structure of the nitrosophenols.   This reaction is a modification of Liebermann's test for nitroso compounds and is given by phenols that have their ortho or para positions unsubstituted so that the quinoid reaction can take place.

It is possible to isolate the reaction product by neutralization of the acid, solvent extraction of the organic from the salts, and subsequent treatment with trimethylsilylation reagents.   Chromatographic examination can separate a variety of phenols as trimethylsilyl derivatives of nitroso compounds (see Fig. 5-35).

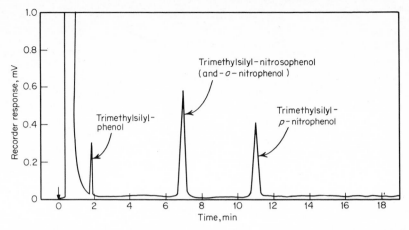

Fig. 5-35 Separation of trimethylsilyl phenol and nitrosophenol on W-98 (10%) column (6 ft × ¼ in.) at 150°C.

Millon's Test

Add 50 mg of the unknown phenol to 1 ml of Millon's reagent (dissolve 10 g of mercury in 7.5 ml of concentrated nitric acid; dilute with 13.5 ml of distilled water).   Place the tube with the sample and reagent in a hot-water bath; heat to boiling.   A red color will be a positive test for a monohydroxyphenol that has at least one ortho position open.   Tyrosine, tyrosine-containing proteins, phenolic acids, and other compounds that have one phenolic group with at least one of its ortho positions open give a positive reaction.

As with previous tests, it is possible to isolate the reaction product by neutralization of the acid and separation of the organic by solvent extraction.   The product may be chromatographed as is or converted into more volatile derivatives (see Fig. 5-36).

Potassium Permanganate

Phenols reduce potassium permanganate solution (2%) and thus undergo oxidation to quinones.   These quinones can be further oxidized with excess potassium permanganate reagent to yield a series of oxidation-reaction products, including maleic acid, oxalic acid, and finally carbon dioxide:

$$\text{(phenol)} \xrightarrow{(O)} \text{(quinone)} \xrightarrow{(O)} \begin{array}{c} \text{CHCOOH} \\ \| \\ \text{CHCOOH} \end{array} \xrightarrow{(O)} \begin{array}{c} \text{COOH} \\ | \\ \text{COOH} \end{array} \xrightarrow{(O)} 6\,CO_2$$

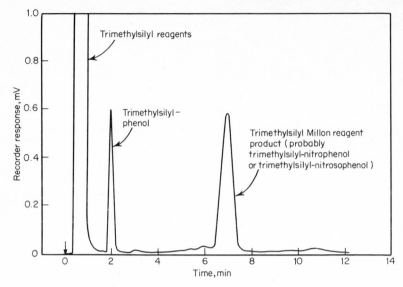

Fig. 5-36 Separation of reaction products of phenols and Millon's reagent on SE-30 (10%) column (6 ft × ¼ in.) at 150°C.

Each stage of the oxidation can be followed by isolation of the products and their conversion into volatile derivatives.

Sodium-Metal Reaction

Add thin slices of clean, freshly cut metallic sodium to a phenol in an inert solvent.   The reaction liberates bubbles of hydrogen as the sodium replaces the active hydrogen on the phenolic group.   However, the phenolic group may be so sterically hindered that it reacts very slowly or not at all.   The solvent and the unknown phenol must be very dry or the bubbling from the reaction of contained water may be mistaken for a reaction with the sodium.

It must be noted that *any* active hydrogen will release bubbles of hydrogen.   Excess unreacted sodium metal is removed and an alkyl chloride cautiously added.   An alkyl aryl ether is produced if the active hydrogen substance is a phenol; and alkyl hydrocarbon is produced if the substance is an alkyl hydrocarbon with an active hydrogen.   Chromatographic examination and comparison with references can assist in the verification of the reaction (see Fig. 5-37).

Ferric Chloride Test

To a 0.1% solution of the unknown phenol in distilled water add a few drops of ferric chloride solution (1%).   A red color is a positive test for phenols or enols.

If the unknown is insoluble in water, dissolve 30 mg of the solid (or

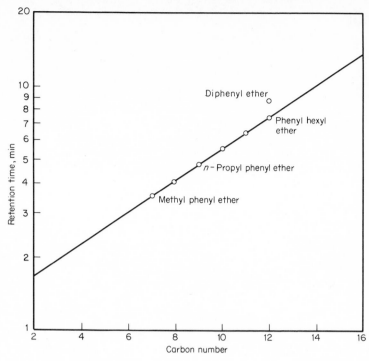

Fig. 5-37   Relationship of aryl ethers to their carbon numbers on W-98 (10%) column (6 ft × ¼ in.) at 150°C.

insoluble liquid) in 1 ml of chloroform and add a few drops of a solution of 0.1 g of ferric chloride in 10 ml of chloroform.   Then add a few drops of pyridine and observe the color.

Phenols and enols produce color with ferric chloride solution.   However, some of the phenols do not produce colors.   A negative test should not be considered significant unless there is other evidence to support the negative results.   Almost all oximes and hydroxamic acids also give a red color with ferric chloride solution (see hydroxamic test under esters).

Compounds Containing Nitrogen in the Functional Group

In this class, halogens or sulfur may also be present in subordinate groups, e.g., halogenated aromatic amines.   Compounds in which both sulfur and nitrogen are present in the same functional group, e.g., a sulfonamide, are discussed under compounds containing sulfur.

AMIDES

Hydroxylamine Test

All common amides give this test. Sterically hindered amides as well as benzanilide and diacetylbenzidine fail to give a positive response to the procedure.[25,42,166]

Add 1 drop of the unknown compound dissolved in propylene glycol to 2 ml of 1 M hydroxylamine hydrochloride, also dissolved in propylene glycol. Add 1 ml of 1 M potassium hydroxide solution. Boil for 2 min and cool. Add 0.5 ml of ferric chloride solution (5%) in alcohol.

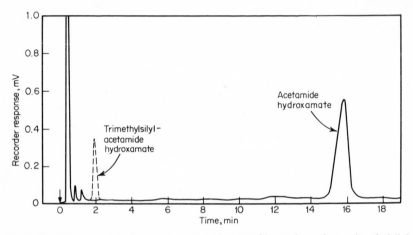

Fig. 5-38 Behavior of hydroxamate ester of acetamide as is and as trimethylsilyl derivative on W-98 (10%) column (6 ft × ¼ in.) programmed from 90 to 250°C at 10°/min.

A red to violet color is considered a positive test. Negative tests are yellow colors; brown colors or precipitates are probably due to complexing or the production of insoluble indeterminate compounds.

$$\underset{RCNH_2}{\overset{O}{\overset{\|}{}}} + H_2NOH \longrightarrow \underset{RCNHOH}{\overset{O}{\overset{\|}{}}} + NH_3$$

The hydroxamate can be further derivatized with trimethylsilyl reagents to produce compounds that will be separable in the chromatograph (Fig. 5-38).

Aliphatic

Hydroxamate Test. Hydroxylamine hydrochloride converts aliphatic amides to hydroxamic acids, which gives a reddish ferric hydroxamate color with ferric chloride solution.   Salicylamide is the only aromatic amide that appears to give this test:

$$RCONH_2 + H_2NOH \cdot HCl \longrightarrow RCONHOH + NH_4Cl$$

To 0.1 g of the unknown amide add 1 ml of hydroxylamine hydrochloride (7 g in 100 ml of methanol).   Boil solution for 3 min with gentle boiling.   Cool and add a drop of ferric chloride solution (10%). A positive test is a wine-red or purple color.

To examine the reaction product, evaporate the solvent, treat with trimethylsilylating reagents, and inject into the chromatograph.   From a homologous series of aliphatic amides one can obtain an accurate estimate of the number of carbons in the alkyl chain (see Fig. 5-39).

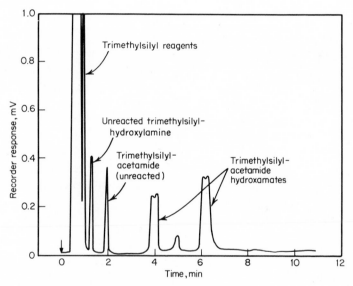

Fig. 5-39   Separation of ammonia and amines on a Carbowax 400 (20%) column (3.75 m × 0.5 cm i.d.) on 80/100 mesh Celite at 80°C.

Hydrolysis.   Sodium hydroxide solution hydrolyzes aliphatic amides readily.   However, aromatic amides and sulfonamides are not hydrolyzed under conditions of the test described below:

$$RCONH_2 + NaOH \longrightarrow RCOONa + NH_3$$

Boil 0.2 g of the unknown with 5 ml of aqueous sodium hydroxide (10%) for 3 min.   Test for ammonia release with moist red litmus paper or absorption in a neutral copper(II) salt solution.   A blue color develops in the presence of ammonia.   The ammonia test described under ammonium salts may also be applied. Chromatographically, ammonia can be detected by injecting the condensate from the caustic treatment into the instrument under the conditions described in Fig. 5-40. This procedure will also help to discriminate between ammonia, amines, and related alkaline volatile compounds. Some

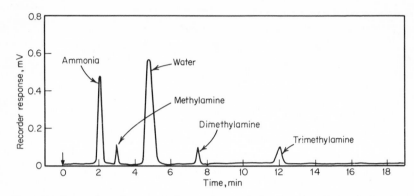

Fig. 5-40 Separation of ammonia from other methylamines on a Poropak Q column (6 ft × ⅛ in.) at 74°C.

classes of compounds that may interfere are the aliphatic amine salts, hydrazines, and some aromatic amines.

If the sodium hydroxide solution is acidified, the free acid may be isolated and chromatographed as is or converted into esters.

Aromatic

Hydroxamate Test. Hydrogen peroxide oxidizes aromatic amides (not sulfonamides) directly to hydroxamic acids in the presence of catalytic amounts of ferric chloride:

$$ArCONH_2 + H_2O_2 \longrightarrow ArCONHOH + H_2O$$

Thus, the hydroxamic acid test is given directly. Mix 50 mg of the powdered amide (or liquid) with 5 ml of distilled water. Add a drop of ferric chloride solution (10%) and ½ ml of hydrogen peroxide (3%). Slowly heat the solution to boiling. The characteristic wine-red color will develop. Heating too long may destroy the color. All organic solvents should be absent.

Soda Lime. If the unknown amide is mixed with soda lime and the mixture is dry-fused, nearly all of the simple amides will be decomposed. Thus aliphatic amides, aromatic amides, and sulfonamides, as well as the *N*-substituted amides, i.e., anilides and *N*-dimethyl aromatic amides, will be decomposed. The simple amides yield ammonia, and the *N*-substituted amides yield primary or secondary amines:

$$RCONH_2 + CaO \cdot NaOH \longrightarrow RCOONa + NH_3 + CaO$$

$$ArCON(R)_2 + CaO \cdot NaOH \longrightarrow ArCOONa + CaO + R_2NH$$

$$RCONHAr + CaO \cdot NaOH \longrightarrow RCOONa + CaO + ArNH_2$$

$$ArSO_2NH_2 + CaO \cdot NaOH \longrightarrow ArSO_3Na + CaO + NH_3$$

Mix well 0.5 g of the unknown amide with 2 g of dry soda lime and place in a dry flameproof test tube. Attach a distillation tube from the mouth of the tube to the bottom of another cooled test tube containing a few drops of water. Heat the tube containing the soda-lime mixture until the mixture begins to fuse, applying a flame along the tube to move the condensate into the receiver. Apply a hot flame to the reaction mixture for 1 min. Test the gas that is evolved for ammonia. Gas-chromatographic examination is a useful technique under conditions as described (Fig. 5-41). If the tests for aliphatic amides and for sulfon-amides are negative, the compound may be considered to be an aromatic amide. If both ammonia and an amine are found, it is probable that the ammonia was a product of decomposition of an *N*-substituted amide. Complete fusion of the salts remaining in the soda lime with excess sodium hydroxide will form a hydrocarbon and sodium carbonate (or sodium sulfite in the case of the sulfonamide). The distillate should be examined for amines as well as hydrocarbons, both of which may be examined chromatographically, as shown in Fig. 5-42.

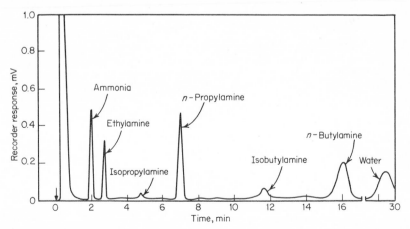

Fig. 5-41 Separation of ammonia and amines on Carbowax 400 (20%) column (3.75 m × 0.5 cm) at 80°C.

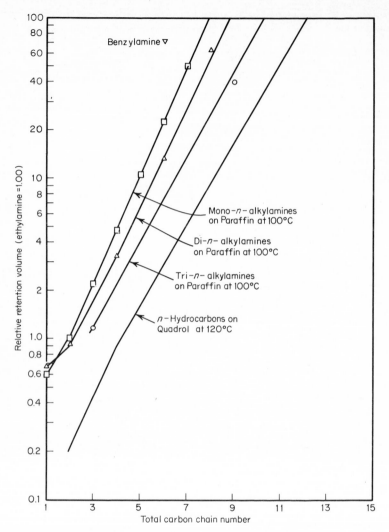

Fig. 5-42 Examination of soda-lime fusion condensate for hydrocarbons and amines separated on paraffin and Quadrol (10%) columns (6 ft × ¼ in.) at 100 and 120°C.

AMMONIUM SALTS

On warming with sodium hydroxide or soda lime, ammonium salts evolve free ammonia. Similarly, reaction of ammonium ions with sodium hypochlorite in the presence of phenol produces a blue color, due to the oxidation of the ammonia to nitrous acid, which then reacts with the phenol to produce a nitrosophenol.   The nitrosophenol is then converted to a quinone by further action of the hypochlorite.

Mix 1 ml of phenol in water (4%) with 1 ml of aqueous sodium hypochlorite (5%) in a test tube.   Add a few crystals or a few drops of an ammonium salt and warm the mixture.   A blue color is a positive test for the presence of an ammonium ion.   As with the nitrous test for phenol, this compound may be isolated, treated with trimethyl-silylating agents, and chromatographed, as shown in Fig. 5-43.

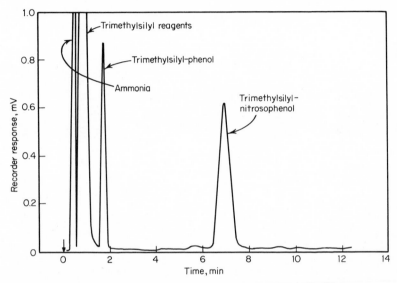

Fig. 5-43   Separation of nitrosophenol from ammonia as the trimethylsilyl derivatives on W-98 (10%) column (6 ft × ¼ in.) at 150°C.

ANILIDES

Anilides decompose into free aniline when fused with soda lime.   The aniline can be identified by the tests described under amines below or chromatographed under the conditions shown in Fig. 5-42.   The anilides also hydrolyze sufficiently to give a positive test for aniline by use of the carbylamine reaction (isocyanide formation).   Almost all the anilides produce a rose or purple color with concentrated sulfuric acid and potassium dichromate (Tafel's test).   Shake 0.1 g of the unknown compound with 3 ml of concentrated sulfuric acid and add 50 mg of powdered potassium dichromate; do not heat.   A rose or purple color is positive for anilides.   Anilines and anilides are detectable by gas chromatography (see Fig. 5-44).

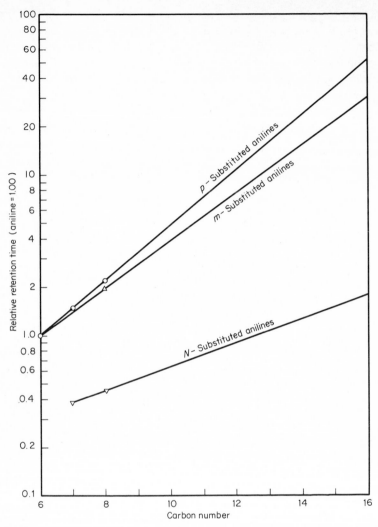

Fig. 5-44 Separation of anilines and anilides on Apiezon L (5%) column (3 ft × ¼ in.) at 150°C.

AMINES

Unfortunately there is no one test that is specific for, or works on, all amines. The primary amines are easily identified, but many of the substituted aromatic amines (even some primary aromatic amines) do not dissolve in dilute acids. It is more difficult to obtain satisfactory identification of the polyaryl amines by the usual tests. However, gas chromatography plus other instrumentation greatly simplifies the identification.   Since a single specific test for all amines cannot be relied on, it is better to subject any nitrogenous compound to at least two or three other amine tests.

AMINE SALTS

Using aqueous sodium hydroxide (10%), neutralize the solution of the amine salt.   Distill off the free amine, or extract it with a solvent such as ether.   Run several of the appropriate tests described below on the free amine.   Inject into the chromatograph.

Hinsberg Reaction (Benzenesulfonyl Chloride)

This procedure is used to distinguish between primary, secondary, and tertiary amines.   The benzenesulfonyl chloride reacts with primary and secondary amines but does not react with tertiary amines.   The benzenesulfonamide subsequently formed is further reacted with dilute aqueous sodium hydroxide solution.   Primary amides dissolve in sodium hydroxide, while secondary amides are not soluble in the alkali unless they are amphoteric, like an alkyl amino acid or phenol.   Some substituted amines do not respond to this test:

Primary: $SO_2Cl + RNH_2 + 2NaOH \longrightarrow$

$$Na^+ \; \overset{(-)}{\left[\underset{}{\bigcirc} - SO_2NR \right]} + NaCl + 2H_2O$$

Soluble

Secondary: $SO_2Cl + R_2NH + NaOH \longrightarrow$

$$\bigcirc - SO_2NR_2 + NaCl + H_2O$$

Insoluble

Tertiary: $SO_2Cl + R_3N + NaOH \longrightarrow$ no reaction

Place 0.2 g of the amine, 0.3 ml of benzenesulfonyl chloride, and 10 ml of sodium hydroxide (5%) in a test tube.   Stopper and shake vigorously for 2 min.   Cool under running cold water if the reaction heats.   Check the acidity or basicity with litmus.   Add more alkali and shake again for acid solutions.   If all of the original compound dissolves, the unknown is a primary amine.   When a residue remains (liquid or solid), test its solubility in hydrochloric acid (25%).   If the residue is soluble in the acid, the unknown probably is a tertiary amine.   If the residue does not dissolve in the acid, the unknown is a secondary amine that has reacted with the reagents.   Purify the residue and measure its melting point.

If the amines are mixed and it is desirable to separate them, hydrolyze the benzenesulfonamides by heating 10 g of these reaction products with 100 ml of hydrochloric acid (25%) and reflux.   The benzenesulfonamides of the primary amines require 24 to 36 hr of refluxing.   The secondary benzenesulfonamides hydrolyze in 10 to 12 hr.   The solution may be cooled and alkalinized with aqueous sodium hydroxide (20%) and the amine extracted with 3 to 5-ml portions of inert solvent such as ether.   The separated amines can be derivatized further or examined by the chromatograph as is (see Fig. 5-45).

Acetyl Chloride Reaction

Acetyl chloride is a very reactive reagent, reacting with water, alcohols, and other compounds having replaceable hydrogens.   This includes many of the primary and secondary amines.   The aromatic amines are much less reactive; however, some of the substituted aromatic amines do not appear to react with the reagent.   The nitro groups, particularly in the ortho or para positions, inactivate the amino group, making the amine unable to dissolve in dilute hydrochloric acid.   When these amines react with the acetyl chloride reagent, N-substituted amides are formed:

$$CH_3-\!\!\left\langle\bigcirc\right\rangle\!\!-NH_2 + CH_3COCl \longrightarrow$$

$$CH_3-\!\!\left\langle\bigcirc\right\rangle\!\!-NHCOCH_3 + HCl$$

$$(C_2H_5)_2NH + CH_3COCl \longrightarrow (C_2H_5)_2NCOCH_3 + HCl$$

Place 0.1 g of the amine under test in a tube and add 2 drops of acetyl chloride.   If no reaction occurs (as evidenced by no heat production), warm the tube.   The amide will then form and can be purified for melting-point measurement or chromatographed as shown in Fig. 5-46.

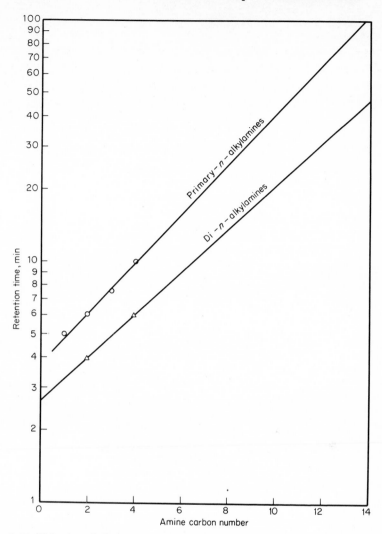

Fig. 5-45   Behavior of alkyl amines as *p*-toluenesulfonamides on SE-30 (1%) column (12 ft × 4 mm) at 175°C.

Primary Aryl Amines

Diazotization Test. Substituted amines that are insoluble in dilute hydrochloric acid and do not react with acetyl chloride usually give the diazotization test.   Nitrous acid converts primary aryl amines into diazonium salts, which are fairly stable at low temperatures.   These salts can be coupled to substances that will produce characteristic dye colors, e.g., the sodium salt of the β-naphthol.

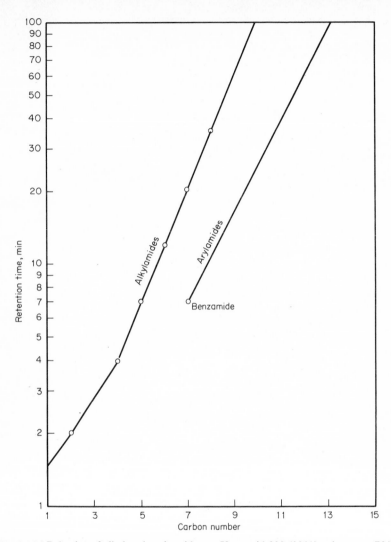

Fig. 5-46 Behavior of alkyl and aryl amides on Versamid 900 (10%) column on 70/80 mesh Chromosorb W, silanized (3 ft × ⅛ in.) at 150°C.

Into a cold bath of 100 ml of crushed ice, salt, and water place a tube of 1 ml of water, 5 drops of concentrated sulfuric acid, and 0.1 g of the unknown primary aryl amine.   In another tube add 1 ml of sodium nitrite solution (10%) and chill in the ice bath.   Using a third chilled tube, add 2 ml of sodium hydroxide solution (10%) containing 0.2 g of dissolved β-naphthol.   Add the sodium nitrite dropwise to the acidified aryl amine with shaking. Then add the alkaline β-naphthol in drops. The presence of primary aryl amine is indicated by a red color. If different monohydroxyphenols are used, the colors may be other than red.

$$\text{C}_6\text{H}_5\text{—NH}_2 + \text{HONO} + \text{H}_2\text{SO}_4 \longrightarrow \text{C}_6\text{H}_5\text{—N}_2\text{HSO}_4 + 2\text{HOH}$$

$$\text{C}_6\text{H}_5\text{—N}=\text{NHSO}_4 + \underset{\text{ONa}}{\text{naphthol}} + \text{NaOH} \longrightarrow$$

$$\underset{\text{ONa}}{\text{naphthol—N}=\text{N—C}_6\text{H}_5} + \text{NaHSO}_4 + \text{H}_2\text{O}$$

To chromatograph the final dye product, acidify; extract the product with an inert solvent such as ether; dry; and treat with trimethyl-silylating reagents (see Fig. 5-47).

Carbylamine Test. Almost all primary amines react with chloroform and potassium hydroxide to form the isocyanides, which are irritatingly

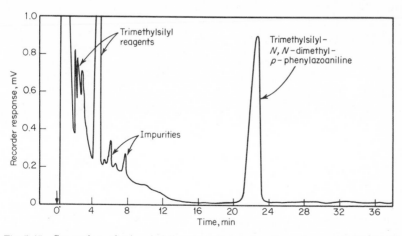

Fig. 5-47 Separation of trimethylsilyl derivatives of a diazo dye on W-98 (10%) column (6 ft × ⅛ in.) at 250°C.

odorous and may be toxic.   This procedure will also detect traces of primary amines in other amines.   It will similarly detect primary aryl amines:

$$C_3H_7NH_2 + CHCl_3 + 3KOH \longrightarrow C_3H_7NC + 3KCl + 3H_2O$$

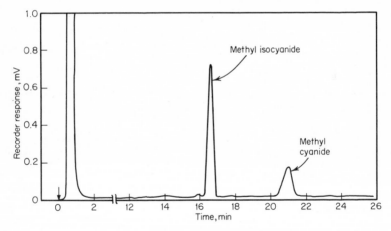

Mix 50 mg of the unknown primary amine with 2 drops of chloroform and 1 ml of 2 N potassium hydroxide in methanol.   **Cautiously** warm and **cautiously** note odor.   Inject portions of the reaction mixture into the chromatograph and note the location of peaks, as shown in Fig. 5-48.

Fig. 5-48   Examination of isocyanides on Apiezon L (5%) column (3 ft × ⅛ in.) at 150°C.

Nitro- or Nitroso-Substituted Aryl Amines.   If an aryl amine with a nitro or nitroso group in the ortho or para position relative to the amine is boiled with sodium hydroxide solution (10%), the amine is converted into a substituted phenol.   Primary aryl amines give a positive test, and so do the n-alkyl substituted amines.   Ammonia or an alkyl amine is released.   The amines can be collected in the distillate and examined chromatographically (see under amines).

The color of the solution changes from yellow to orange or red, due probably to a quinone-type structure:

$$(CH_3)_2N\!-\!\!\langle\bigcirc\rangle\!-\!NO + NaOH \longrightarrow$$

$$(CH_3)_2NH + NaO\!-\!\!\langle\bigcirc\rangle\!-\!NO$$

$$NaO\!-\!\!\langle\bigcirc\rangle\!-\!NO \rightleftharpoons O\!=\!\!\langle\bigcirc\rangle\!=\!NONa$$

$$H_2N\!-\!\!\langle\bigcirc\rangle\!-\!NO_2 + NaOH \longrightarrow NH_3 + NaO\!-\!\!\langle\bigcirc\rangle\!-\!NO_2$$

$$NaO\!-\!\!\langle\bigcirc\rangle\!-\!NO_2 \rightleftharpoons O\!=\!\!\langle\bigcirc\rangle\!=\!NOONa$$

In a test tube equipped with a distilling tube dipping into a cooled tube of water heat 0.1 g of the unknown amine with 5 ml of sodium hydroxide (10%) until the solution boils. Test vapors for ammonia or amine and note the color change in the solution.

The phenol may be acidified, shaken out with an inert solvent, converted into trimethylsilyl derivatives, and chromatographed (see under phenols).

PRIMARY AND SECONDARY ALKYL AMINES

The amines should be water-soluble for these tests. Mixtures of both types of amines will give positive results, though the test for a secondary amine is masked by excess of the primary amine.

Primary

To 5 ml of a dilute water solution of the amine add 1 ml of acetone (aldehyde-free) and 1 drop of sodium nitroprusside solution (10%). A violet-red color developing in 1 min is a positive test for a primary amine.

Secondary

To 5 ml of a dilute water solution of the amine, add 1 ml of acetaldehyde solution (5%) and then add 1 drop of sodium nitroprusside solution (1%). A blue color developing within 5 min is a positive test for a secondary amine. On standing the color changes to green and then yellow.

Chloranil Test (Tetrachloroquinone). Add 3 drops of a saturated solution of chloranil in dioxane to 1 drop of the unknown amine. Red, blue, or green colors develop with most amines. Some phenols and divalent sulfur-containing compounds give amber or orange colors.

Most of these compounds have not been gas-chromatographed satisfactorily, probably due to decomposition.

Schotten-Baumann Reaction. Aroyl chlorides or acid chlorides react with amines in the presence of sodium hydroxide to form aroyl or acyl derivatives without the formation of other products. Alcohols may be

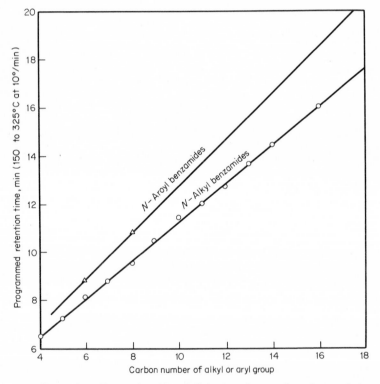

Fig. 5-49 Separation of benzoyl amides of aliphatic and aromatic amines on Apiezon L (5% and 3% KOH) column (3 ft × ⅛ in.) at 175°C.

esterified competitively and should therefore be absent:

$$
\underset{RC-Cl}{\overset{O}{\overset{\|}{}}} + R'OH + NaOH \longrightarrow \underset{RC-OR'}{\overset{O}{\overset{\|}{}}} + NaCl + H_2O
$$

$$
\underset{ArC-Cl}{\overset{O}{\overset{\|}{}}} + R'NH_2 + NaOH \longrightarrow \underset{ArC-NHR'}{\overset{O}{\overset{\|}{}}} + NaCl + H_2O
$$

Drop 1 ml of the unknown amine into a glass-stoppered flask containing 5 ml of water, 1 ml of benzoyl chloride, and 5 ml of aqueous sodium hydroxide solution (20%). Agitate the mixture by shaking. Test the solution with litmus paper to ascertain that the reaction is still alkaline.[86,169] The amide can be purified and tested further or chromatographed, as shown in Fig. 5-49.

SECONDARY AMINES

Nickel–Carbon Disulfide Complex Reagent

Add enough carbon disulfide to 0.5 g of nickel chloride, $NiCl_2 \cdot 6H_2O$, in 100 ml of distilled water for a droplet to remain on top of the water after shaking. Keep tightly stoppered, replacing carbon disulfide as it evaporates.

Add several drops of the amines to 5 ml of distilled water; then add 1 to 2 drops of concentrated hydrochloric acid, if necessary, to solubilize the amine. To 1 ml of the reagent above, add 0.5 to 1 ml of the amine solution. A definite precipitate is a positive test for a *secondary* amine. A slight turbidity indicates that a secondary amine is present only as an impurity. All secondary amines respond to this test. Primary or tertiary amines *do not* respond in this manner (except for turbidities due to impurities). Substituted pyridines, quinolines, and isoquinolines also respond:[43]

$$
R_2NH + CS_2 + NH_4OH \longrightarrow \underset{R_2N-C-SNH_4}{\overset{S}{\overset{\|}{}}} + H_2O
$$

$$
\left[\underset{R_2N-C-S}{\overset{S}{\overset{\|}{}}} \right]_2 Ni \xleftarrow{\quad\quad} \Big|^{NiCl_2}
$$

The complex decomposes in the injection port of the chromatograph, yielding carbon disulfide and the free amine, as shown in Fig. 5-50.

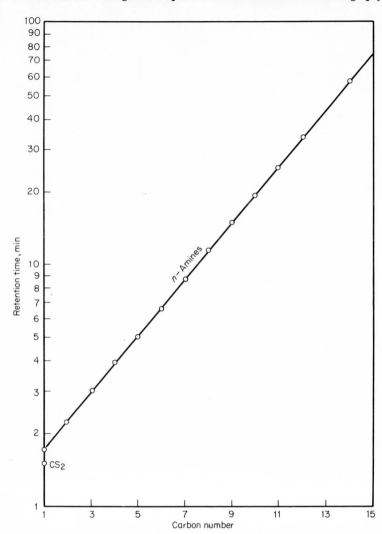

Fig. 5-50   Decomposition of nickel complex to yield the amines and carbon disulfide separated on Chromosorb 103 column (3 ft × ¼ in.) at 85°C.

PRIMARY ALIPHATIC AND AROMATIC AMINES

Nickel 5-Nitrosalicylaldehyde Reagent

To 15 ml of triethanolamine add 0.5 g of 5-nitrosalicylaldehyde and 25 ml of distilled water.   If the triethanolamine is contaminated with monoethanolamine, add another 0.5 g of 5-nitrosalicylaldehyde and remove the impurity precipitate by filtration.

Solubilize the amine by adding a few drops of it to 5 ml of distilled water. It may be necessary to add a few drops of concentrated hydrochloric acid to bring the amine into complete solution. Add 0.5 ml of this solution to 3 ml of the test reagent. Primary aliphatic amines produce an immediate voluminous precipitate. Aromatic primary amines take 2 to 3 min to produce a distinct precipitate. A turbidity indicates only impurities.

All primary amines respond to this test. Hydroxylamines and hydrazines (monosubstituted on only one hydrogen) give a definite precipitate. Amides *do not* give a positive test, and amino acids are not reactive to this test:

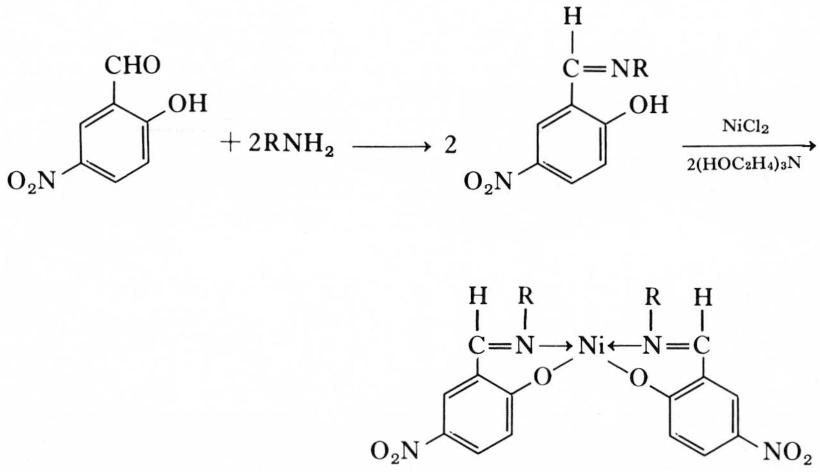

This complex is usually too large a molecule and too unstable to be chromatographed satisfactorily unless it is protected against decomposition.

Sodium-Metal Reaction

Sodium metal reacts with easily replaceable hydrogens:

$$2R_2NH + 2Na \longrightarrow 2R_2NNa + H_2$$
$$RNH_2 + 2Na \longrightarrow RNNa_2 + H_2$$

Care must be exercised with this test, as traces of moisture will cause bubbling that may be interpreted as reaction with the compound under test.

Thus primary and secondary amines will react, but tertiary amines will not. Since any reactive hydrogen in the molecule will show a positive test, care must be exercised in its interpretation. Hydroxyl groups and hydrogens on adjacent unsaturated bonds and adjacent to activating groups will react.

Treatment of the product (after removal of excess sodium) with various derivatizing agents converts the product into other compounds that can be examined by other physical or chemical means or by the gas chromatograph.

HYDRAZINES

The hydrazines decompose when heated with aqueous sodium hydroxide (10%), yielding ammonia.   They also reduce Tollen's and Benedict's reagents.   Hydrazones are formed by reaction with aldehydes and ketones.

Hydrazines as a whole chromatograph without appreciable decomposition if the temperatures on the injection port, column oven, and detector are not above the decomposition temperature (see Fig. 5-51). Some of the hydrazones can be chromatographed satisfactorily (see Fig. 5-52).

To prepare a hydrazone, suspend 0.1 ml of the unknown hydrazine in 1 ml of distilled water.   Add a few drops of glacial acetic acid to solubilize the unknown hydrazine, then a few drops of a 5% solution of an aldehyde such as acetaldehyde or propionaldehyde.   A precipitate will form if the substance is an unsubstituted or partially substituted hydrazine.   Check the melting point and inject into a chromatograph. Compare melting point and retention time with a known hydrazine and aldehyde converted to hydrazone.   Also check retention time of the original unreacted hydrazine with a known sample (see Fig. 5-52).

Dinitrophenylhydrazines are usually highly colored, forming insoluble

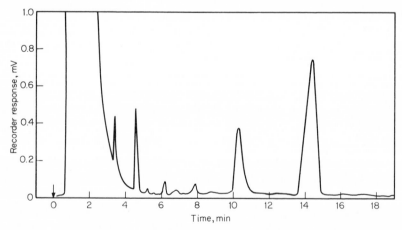

Fig. 5-51 Separation of hydrazines as trimethylsilyl derivatives on Chromosorb 103 column (3 ft × ⅛ in.) at 70°C (injector and detector must be kept below 100°C).

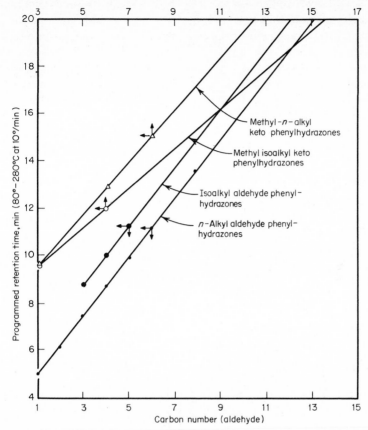

Fig. 5-52 Separation of hydrazones of aldehydes and ketones on W-98 (10%) column at 100°C (6 ft × ¼ in.).

solids with most aldehydes and ketones. At first some of the solids may be oily, changing to crystalline solids on standing or on further purification. A few fail to form solid derivatives at all, such as di(n-amyl) ketone, methyl n-octyl ketone, and others.

Occasionally, certain alcohols may be oxidized by the 2,4-dinitrophenylhydrazine to aldehydes or ketones, which then would give a positive reaction, e.g., cinnamyl alcohol, 4-phenyl-3-butene-2-ol, vitamin A alcohol, and benzohydrol.[18] Also certain alcohols may be contaminated with enough aldehyde or ketones or be oxidized by the air to give a positive test. If a very small amount of hydrazone seems to be formed, it may be only an impurity or an oxidation product. Check the sample on a gas chromatograph before and after reaction with the hydrazine.

The 2,4-dinitrophenylhydrazones give various colored compounds depending upon the structure of the reacting aldehyde or ketone. Yellow is the usual color of the hydrazone in which the carbonyl group is not conjugated with another functional group. A carbon-carbon double-bond conjugation or a bond with a benzene ring shifts the absorption toward the visible;[153] another absorption is observed in the ultraviolet. This shift changes the color from yellow to orange-red.

Trimethylsilyl reagents react with the hydrazines, producing more volatile compounds that are more easily separated in the instrument (see Fig. 5-53). To prepare these derivatives, add 10 mg of the dry hydrazine to a mixture of 1 ml of pyridine, 0.2 ml of hexamethyldisilazane, and 0.1 ml of trimethylchlorosilane. Shake the mixture vigorously and allow to stand for 5 min. Inject into the chromatograph. Ignore the first three peaks, which are usually the reagents.

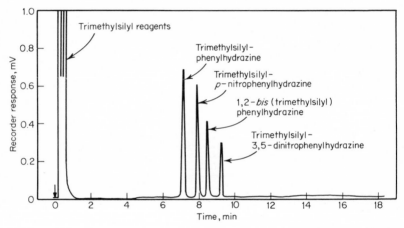

Fig. 5-53 Separation of trimethylsilyl derivatives of hydrazines and phenylhydrazines on W-98 (10%) column (6 ft × ¼ in.) programmed from 100 to 250°C at 10°/min.

Phenylhydrazines produce oils with most aldehydes and ketones. They can be observed by chromatography, as shown in Fig. 5-54. *p*-Nitrophenylhydrazine, however, forms crystalline derivatives with most aldehydes and ketones. To prepare derivatives from these hydrazines, add 0.5 g of the *p*-nitrophenylhydrazine or phenylhydrazine to 5 ml of water; then drop in glacial acetic acid until the hydrazine just dissolves. Add 0.5 ml of an aldehyde or ketone to 2 ml of ethanol and add water until the cloudiness just disappears. Mix the two solutions; if no precipitate or oil appears, warm slightly.

Most hydrazines will form osazones with carbohydrates having

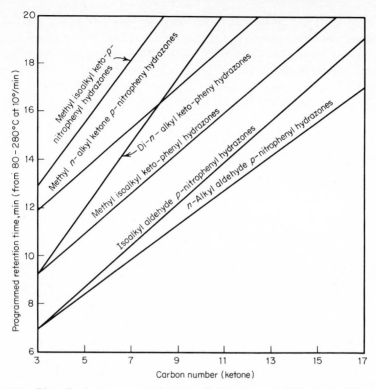

Fig. 5-54 Phenylhydrazones or *p*-nitrophenylhydrazones of aldehydes and ketones separated on W-98 (silicone gum rubber, 10%) column as is and as trimethylsilyl derivatives (6 ft × ¼ in.) from 100 to 250°C at 10°/min.

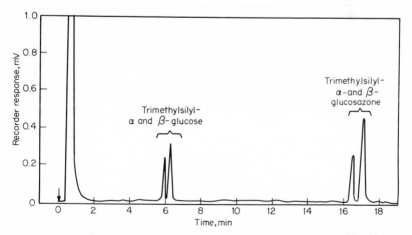

Fig. 5-55 Separation of osazones as trimethylsilyl derivatives on W-98 (silicone gum rubber, 10%) column (6 ft × ¼ in.) from 60 to 250°C at 10°/min.

reactive carbonyl groups. These osazones are not easily chromatographed until the hydroxyl groups are protected by acetate or trimethylsilyl ether formation (see Fig. 5-55). Most osazones crystallize in characteristic forms, which have been observed under the microscope.[133]

NITRO, NITROSO, AZOXY, AND AZO COMPOUNDS

General Test. This procedure will serve only to establish that the compound is one of the above types. Specific tests for each can then be applied to verify which group is present. The compound is reduced to hydroxylamine, hydrazine, or hydrazo compounds, and their further reducing action is tested with Tollens' reagent. The initial reducing action must be tested first. Suspend or dissolve 0.1 g of the unknown compound in 2 ml of ethanol and add 1 ml of Tollens' reagent. Allow to stand 5 min; a silvery coating or black precipitate indicates that the substance has reduced this reagent. Further testing is not necessary.

To reduce the substance, suspend or dissolve 0.2 g of the unknown compound in 3 ml of hot ethanol (50%). Add 3 drops of glacial acetic acid and about 0.1 g of zinc dust. Heat to boiling and allow to cool 5 min. Filter and place two halves of the filtrate in separate test tubes. To one half of the filtrate add 2 ml of Tollens' reagent and allow the solution to stand for 5 min. If a silvery mirror or a black precipitate forms, the compound has a nitro, nitroso, azoxy, or azo group. Also test a portion by the chromatograph.

If the above test is positive, add 1 drop of benzoyl chloride to the second portion of the filtrate and warm the mixture. Add 1 drop of concentrated hydrochloric acid and 1 drop of ferric chloride reagent (10%). A wine-red color (of ferric hydroxamate) indicates that the original compound was a nitro or nitroso compound. Examine the hydroxamate ester by the chromatograph (see esters).

Zinc and Ammonium Chloride Test. Dissolve 0.5 g of a nitro aromatic compound in 10 ml of ethanol (50%). Add 0.5 g each of ammonium chloride and zinc dust. Shake and heat to boiling. Cool 5 min and filter. Test filtrate with Tollens' reagent:

$$\langle\!\langle\bigcirc\rangle\!\rangle\!-\!NO_2 + 4H \xrightarrow[\text{NH}_4\text{Cl}]{\text{Zn}} \langle\!\langle\bigcirc\rangle\!\rangle\!-\!NHOH + H_2O$$

Lithium aluminum hydride will reduce nitro compounds to azo compounds.

Nitro Compounds

Ferrous Hydroxide Oxidation Test. Hearon and Gustavson[90] have shown that nitro compounds will oxidize ferrous hydroxide to the ferric state. The reagent changes color from green to brown. Nitroso compounds, quinones, hydroxylamines, alkyl nitrates, and alkyl nitrites also oxidize ferrous hydroxide. Other tests can be utilized to differentiate between these compounds and nitro compounds.

$$RNO_2 + 4H_2O + 6Fe(OH)_2 \longrightarrow RNH_2 + 6Fe(OH)_3$$

In a 3-in. test tube mix 20 mg of the unknown nitro compound with 1.5 ml of freshly prepared solution of ferrous ammonium sulfate (5%). Bubble natural gas through the tube to remove the air. Add 1 drop of 3 M sulfuric acid and 1 ml of 2 N potassium hydroxide in ethanol or methanol. Quickly stopper the tube and shake vigorously. Note the color of the precipitate: a brown color is a positive test. Inject a small portion of the clear supernatant liquid into the chromatograph and compare retention time of the original with the reduced compound, as shown in Fig. 5-56.

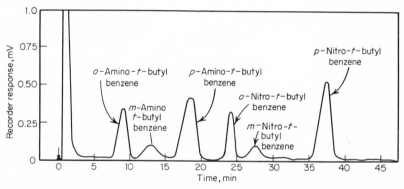

Fig. 5-56 Gas-chromatographic comparison of original and reduced nitro compounds on Chromosorb 103 column (3 ft × ⅛ in.) from 60 to 250°C at 10°/min.

Nearly all nitro compounds give a positive brown color within 30 sec. The speed of the reaction depends largely on its solubility in the reagents. For example, 1-nitronaphthalene is quite insoluble but will give the test in 30 sec. However, 3-nitrophthalic acid is quite soluble in the reagent and will give the positive test immediately.

The Polynitro Compounds (Di, Tri, etc.)

If the above test is positive, Bost and Nicholson[17] have given a test that is useful for estimating the number of nitro groups present in various benzene compounds. A mixture of acetone and sodium hydroxide is used as reagent. The mononitro compounds do not produce appreciable color changes; dinitro compounds produce a purplish-blue color; and trinitro compounds produce deep red colors. However, it must be pointed out that the presence of amino, alkylamino, acylamino, hydroxy, or acylated hydroxy groups on the benzene ring interferes with the test (see Table 5-1).

TABLE 5-1   Exceptions to Alkaline Test

Compound	Color
3,5-Dinitrosalicylic acid	Yellow
2,4-Dinitrophenol	Yellow-orange
2,4-Dinitroresorcinol	Brownish-green
2,4-Dinitroacetanilide	Reddish-orange
1,4-Dinitrobenzene	Greenish-yellow
1,2-Dinitrobenzene	No exception
2,4-Dinitroaniline	Red

Add 0.1 g of the unknown compound to 10 ml of acetone, and then add 3 ml of sodium hydroxide solution (5%). Shake the tube well. The color develops rapidly. The original and the alkaline solutions may be examined on the spectrophotometer. The color shift can be shown in the spectrophotometric traces of the substances. Each is characteristic for a given compound. Ultraviolet scans are also useful in identification.

Nitro Paraffins

The nitrated aliphatic hydrocarbons (nitro paraffins) can be reduced to hydroxylamine derivatives by zinc dust and acetic acid. The filtered solution will reduce Tollens' reagent and will react with benzoyl chloride, hydrochloric acid, and ferric chloride to give a wine-red color characteristic of ferric hydroxamate.

Some of the nitroparaffins will oxidize ferrous hydroxide, but not all of them, e.g., nitromethane and 2-nitropropane, give positive tests. However, nitroethane and 1-nitropropane do not oxidize the reagent under the conditions of the test. Only vigorous reduction will effectively convert the relatively inert nitro paraffins into primary alkyl amines. These may be examined by the gas chromatograph.

Nitrous acid is a convenient reagent for differentiating between the primary, secondary, and tertiary nitro paraffins. Primary nitro paraffins produce a reddish-amber color, secondary nitro paraffins give a sky-blue color, and the tertiary nitro paraffins do not yield any appreciable colors. The action of nitrous acid produces nitrolic acid with a primary nitro paraffin. These are red in solution, and they do not chromatograph, being explosive when dry. The action on a secondary nitro paraffin produces pseudonitroles, which are blue in solution. No reaction takes place with a tertiary nitro paraffin.

Primary: $RCH_2NO_2 + HONO \longrightarrow$

$$\begin{bmatrix} & NO \\ & | \\ R-&C-NO_2 \\ & | \\ & H \end{bmatrix} \longrightarrow R-C\overset{\nearrow NOH}{\underset{\searrow NO_2}{}} + HOH$$

Secondary: $R_2CHNO_2 + HONO \longrightarrow R_2C\overset{\nearrow NO}{\underset{\searrow NO_2}{}} + HOH$

Tertiary: $R_3CNO_2 + HONO \longrightarrow$ no reaction

To 2 ml of sodium hydroxide solution (10%) add 5 drops of the nitro paraffin under test and allow it to stand for 3 min. Then add 1 ml of sodium nitrite solution (10%) and (in drops) sulfuric acid solution (10%). Do not add enough to completely neutralize the solution: a pH of 8 or 9 is sufficient. The color should develop fairly rapidly.

Nitrosoamines

A modification of Liebermann's test (in which phenol is reacted in sulfuric acid with the nitrosoamine), this test is not generally given by nitroso groups attached to carbon atoms but only by nitroso groups attached to nitrogen. The sulfuric acid reacting on the nitroso group produces nitrous acid, which then reacts with the phenol, producing nitrosophenol. The blue color developed is probably due to the monomer of nitrosophenol (the dimeric polymer is colorless); the red color is probably due to quinone monoxime.[32]

Dissolve 0.1 g of phenol in 1 ml of cold concentrated sulfuric acid. Add 50 mg of the nitroso compound under test. Shake the tube and warm slightly by running warm water over the bottom. The blue color will develop if a nitrosoamine is present. Pour the solution slowly into 5 ml of cold water; the color should change to red.

Most nitrosoamines can be chromatographed without appreciable decomposition if instrumental conditions are not excessive. They can also be converted into various derivatives that are stable (see Fig. 5-57).

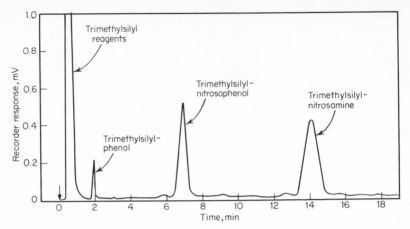

Fig. 5-57 Separation of nitrosamines as trimethylsilyl derivatives on W-98 column (10%) (6 ft × ¼ in.) from 60 to 250° at 10°/min.

ALKYL NITRITES

Caution: Most alkyl nitrites are powerful physiological agents and have strong actions on the heart even at low concentrations. Great care should be taken in handling these substances: avoid breathing the vapors and contact with skin.

Alkyl nitrites can be detected by their ability to react with phenylindole to precipitate 3-isonitroso-2-phenylindole:

Dissolve 0.1 g of 2-phenylindole in hot ethanol and add 0.1 g of the alkyl nitrite under test. Cool the solution. 3-Isonitroso-2-phenylindole should precipitate if an alkyl nitrite is present. The precipitate can be recrystallized from an ester solvent, such as *n*-amyl acetate, and the melting point determined (280°C). The phenylindole can also be chromatographed, but it should be protected against decomposition and converted into a more volatile compound by producing the trimethylsilyl derivatives (see Fig. 5-58).

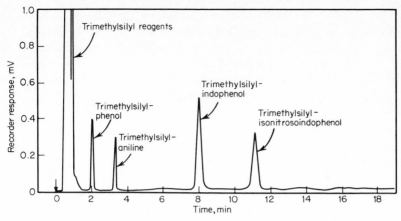

Fig. 5-58 Separation of phenylindole and isonitrosophenylindole as trimethylsilyl derivative on W-98 column (10%) (6 ft × ¼ in.) from 60 to 250°C at 10°/min.

NITRATES

The use of diphenylamine in sulfuric acid is a sensitive test for nitrates. However, both nitrates and nitrites are capable of oxidizing diphenyl-amine reagent.   Most organic nitrates are quite inert, but enough free nitrates are liberated by the sulfuric acid to give the test.   The diphenyl-amine is oxidized to diphenylbenzidine and then to the quinoid (colored) form:

$$2 \quad \text{Diphenylamine} \quad \xrightarrow[\text{NO}_3]{\text{H}_2\text{SO}_4}$$

Diphenylamine

Diphenylbenzidine

Diphenylbenzidine quinone

Add 100 mg of the unknown compound to 3 ml of diphenylamine reagent (0.2 g in 100 ml of concentrated sulfuric acid).   A blue color is a positive indication that nitrates are present.

The diphenylbenzidine quinone is usually too nonvolatile to be effectively gas-chromatographed. Most alkyl or aryl nitrates are easily decomposed in the injection port of the instrument. It is probably more convenient to reduce the nitrates to amines and examine them as compounds or convert to their trimethylsilyl derivatives. Some investigators have analyzed such substances as ethylene glycol dinitrate, propylene glycol dinitrate, etc.[29]

CYANIDES (NITRILES)

Nitriles, especially the cyanohydrins, are hydrolyzed by acids. The nitriles are converted into amides when treated with 90 to 95% sulfuric or concentrated hydrochloric acid at temperatures from 10 to 50°C. By diluting the mixture with water and heating under reflux for ½ to 2 hr the amides can be further hydrolyzed:

$$RCN \xrightarrow[H_2SO_4]{H_2O} RCONH_2 \xrightarrow[H_2SO_4]{H_2O} RCOOH + NH_4HSO_4$$

Nitriles can also be hydrolyzed by boiling with sodium hydroxide solution:

$$RCN + NaOH + H_2O \longrightarrow RCOONa + NH_3$$

Treat 0.2 g of the nitrile with 5 ml of sodium hydroxide (10%) in a test tube. Shake vigorously; then heat to boiling and note whether ammonia is evolved. Acidify and recover the liberated carboxylic

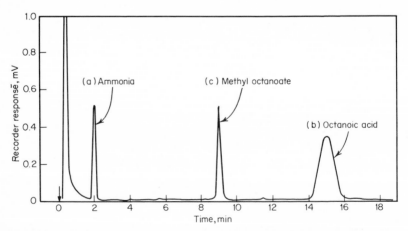

Fig. 5-59 Separation of free ammonia and hydrolysis products on Chromosorb 103 column (3 ft × ⅛ in.) 60 to 250°C at 10°/min; separation of (*a*) ammonia from nitrile, (*b*) free acid, and (*c*) ester.

acid formed. The acid may be chromatographed as is on a polymer-bead column or converted into an ester.

The water solution of the condensate from the hydrolysis may be chromatographed for free ammonia (see Fig. 5-59). The chromatogram of the original nitrile, the free acid, and an ester derivative can be compared with known compounds. Verification can then be made by conversion into other derivatives.

Nitriles also react with hydroxylamine in the presence of potassium hydroxide and ethylene glycol to give a positive hydroxamate test with ferric chloride solution.[42,161]

OXIMES, HYDRAZONES, AND SEMICARBAZONES

Hot concentrated hydrochloric acid on refluxing will convert oximes into hydroxylamine salts, hydrazones into hydrazine salts, and semi-carbazones into semicarbazide salts:

$$\underset{\text{(phenyl)}}{\bigcirc}\!\!-\overset{\overset{\displaystyle H}{|}}{C}\!=\!NOH + H_2O + HCl \longrightarrow$$

$$\underset{\text{(phenyl)}}{\bigcirc}\!\!-CHO + HONH_2 \cdot HCl$$

$$(C_4H_9)_2C\!=\!NNH\!-\!\underset{\text{(phenyl)}}{\bigcirc} + H_2O + HCl \longrightarrow$$

$$(C_4H_9)_2CO + \underset{\text{(phenyl)}}{\bigcirc}\!\!-NHNH_2 \cdot HCl$$

$$CH_3CH_2CH\!=\!NNHCONH_2 + H_2O + HCl$$

$$CH_3CH_2CHO + H_2NCONHNH_2 \cdot HCl$$

Determine whether hydroxylamine hydrochloride is present in the hydrolysate by means of the ferric hydroxamate test. When hydroxylamine is present, it was formed from the oxime. For a negative test, the hydrochloride salt may be condensed with a known aldehyde, e.g., benzaldehyde, or with a known ketone. Determination of the melting point and chromatographic data will help to identify the hydro-chloride salt (see Fig. 5-60).

To hydrolyze these compounds, place 0.2 g of the unknown in a test tube containing 2 ml of concentrated hydrochloric acid.   Place the tube in a boiling water bath for 5 min; then evaporate the mixture to dryness. Add 5 ml of distilled water and filter the solution.   To test for hydrox-ylamine, add 1 drop of benzoyl chloride to 1 ml of the above solution. Alkalize with alcoholic potassium hydroxide and heat to the boiling point.   Cool and acidify with dropwise addition of hydrochloric acid. A wine-red color upon the addition of a few drops of ferric chloride solution (10%) indicates that hydroxylamine was formed from the original oxime.   The hydroxamate ester before addition of ferric chloride solution can be chromatographed (see Fig. 5-61).   If the result is negative, test the remainder of the solution for hydrazine by forming the hydrazone from a known aldehyde, e.g., benzaldehyde, or ketone, e.g., benzophenone, or the semicarbazones.   Use about 0.1 g of the isolated hydrochlorides dissolved in 10 ml of methanol.   Add the aldehyde or ketone dissolved in methanol to the hydrochloride and heat for 15 min.   Recrystallize and compare the melting point with known benzaldehyde hydrazone or semicarbazone or known benzophenone hydrazone or semicarbazone.   Some of these may also be chromato-graphed (see Fig. 5-62).

Other hydrazine tests may be made as described under hydrazines.

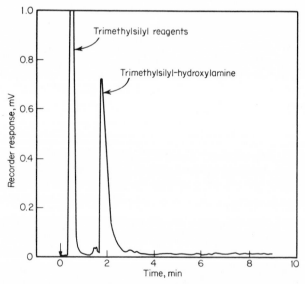

Fig. 5-60 Separation of hydroxylamine as trimethylsilyl derivative on W-98 (10%) column (6 ft × ¼ in.) from 60 to 250°C at 10°/min.

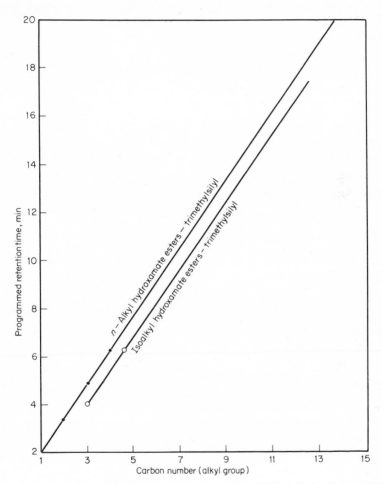

Fig. 5-61 Separation of hydroxamate esters as trimethylsilyl derivatives on W-98 (10%) column (6 ft × ¼ in.) from 60 to 210°C at 10°/min.

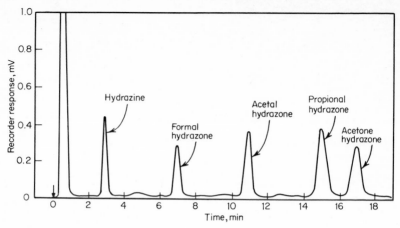

Fig. 5-62 Separation of some aldehyde and ketone hydrazones or semicarbazones as is on W-98 (10%) column (6 ft × ¼ in.) from 60 to 220°C at 10°/min.

Compounds Containing Halogens as Major Functional Group

Many compounds which have halogen as an integral part of their molecules will probably be identified by the presence of other than halogen groups. These latter groups may be more reactive functional groups. Halogenated carboxylic acids, phenols, amines, or sulfonic acids will be identified by their own characteristic group rather than by the less reactive halogens. The presence of halogen will help ascertain the exact identity of the compound, e.g., *p*-chloroaniline, 2,4-dichlorophenoxyacetic acid, *o*-bromophenol, or 2,4,5-trichlorophenoxyacetic acid salt of *n*-bútylamine. The salts of halogenated acids and amines can be identified by treatment with sodium hydroxide solution (the amine can be removed by steam distillation) and the carboxylic acid separated by acidification of the sodium hydroxide solution. Both free amine and free carboxylic acid can be identified by chromatographic examination or other physical and chemical tests or converted into further derivatives.

ARYL, AROYL, AND SULFONYL HALOGENATED COMPOUNDS

The reverse of the Hinsberg test for amines is used, as well as the reverse of the acetyl chloride test. Identification of the unknown is facilitated by using the unknown acyl, aroyl, or sulfonyl halogenated compound and a known aniline as the amine.   Thus acidic halogenated compounds can be identified by their anilide derivatives.   These acidic halogenated compounds can usually be recrystallized from hot water or alcohol-water mixtures.   The chromatogram plus the melting point and other physical properties will assist materially in identifying which specific acidic halogen compound is involved (see Fig. 5-63).

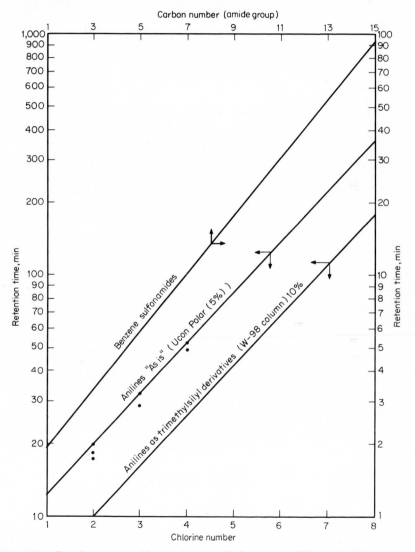

Fig. 5-63   Gas-chromatographic examination of halogenated anilides and sulfonamides as is and as trimethylsilyl derivatives on W-98 (10%) column and Ucon Polar (5%) column (6 ft × ¼ in.) from 80 to 240°C at 5°/min.

To prepare a derivative of these compounds, to 3 drops of a known aniline add 2 drops of the unknown acidic halogen compound. Add 1 ml of cold distilled water. These anilides are sparingly soluble in cold water and should form colorless solids, which can be examined by the chromatograph. The derivative may be purified further and the melting point observed.

Treatment of the acidic halogenated compounds with water should produce the halide acid and a carboxylic or sulfonic acid. Treatment with an alcohol will produce an ester of the alcohol; treatment with pyridine will produce a salt that is easily hydrolyzed in water. Many of these compounds can be observed in the gas chromatograph, or other physical properties can be measured.

Differentiation of Alkyl and Aryl Halides

The halogenated acidic amine salts and the acidic halogen compounds readily form silver halide precipitates with cold aqueous silver nitrate solutions. The aryl and alkyl halogenated substances do not react appreciably with the cold aqueous silver nitrate solution. Alkyl halides give a positive test; however, the aryl halides fail to give the test; exceptions are noted.

Alcoholic Silver Nitrate. Most alkyl bromides and iodides give silver halide precipitates when reacted with alcoholic silver nitrate solution. Benzyl bromide acts like an alkyl halide, as the bromine is on the alkyl group. If the halogen is on the aromatic ring, it fails to give a positive test. A thorough discussion of this reaction has been presented in the literature.[84,162,190]

Add 2 drops of the halogenated compound to 2 ml of alcoholic (ethanol) silver nitrate solution (2%). Aryl halides do not react; alkyl halides will react. The rate indicates the degree of reactivity. The most reactive halogenated acids are those in which the halogens are ionic; amine salts of the halogen acids are the simplest: $RNH_3^+ X^-$.

The oxonium compounds are carbonium salts containing ionic halogens. These uncommon salts are found in crystal-violet dye indicator:

$$\left[R:\overset{..}{\underset{..}{O}}:R\right]^+ X^- \qquad \left[(CH_3)_2N-\!\!\left\langle\!\!\bigcirc\!\!\right\rangle\right]_3 C^+Cl^-$$

Aqueous solutions of silver nitrate give an almost immediate precipitate of silver halides with aqueous solution of oxonium salts.

A summary of the results of the reaction of alcoholic and aqueous silver nitrate solutions with various compounds follows:

A. Water-soluble or hydrolyzable compounds which give instantaneous precipitate with aqueous silver nitrate

  1. Amine salts of halogen acids

  $$(RHN_3)^+X^- + AgNO_3 \longrightarrow AgX + (RNH_3)^+NO_3^-$$

  2. Oxonium salts
  3. Carbonium halides
  4. Low-molecular acid chlorides

  $$RCOCl + H_2O \longrightarrow RCOOH + HCl$$

B. Water-insoluble substances which react with alcoholic silver nitrate solutions

  1. Highly reactive compounds which give an immediate precipitate

  RCOCl   RCHClOR R_3CCl
  RCH=CHCH_2X   RCHBrCH_2Br   RI

  2. Compounds whose reaction is intermediate or slow or nonexistent at room temperature (the latter require heating to give a precipitate)

  RCH_2Cl R_2CHCl $RCHBr_2$

  3. Compounds nonreactive or inert toward cold or hot alcoholic silver nitrate solutions

  ArX   RCH=CHX $CHCl_3$ $ArCOCH_2Cl$
  $ROCH_2CH_2X$

Sodium Iodide in Acetone. This reaction depends upon the slight solubility of sodium chloride and bromide in acetone. Shriner, Fuson, and Curtin[162] have discussed the reactivity and structural relationships. However, it is best to check the reactivity with both reagents since these actions are not identical.

Iodide ion can also take part in an oxidation-reduction reaction, liberating free iodine and forming sodium chloride or bromide.

Dissolve 15 g of sodium iodide in 100 ml of pure acetone. Store in a dark bottle; discard when the solution develops a red-brown color. To 1 ml of the reagent add 2 drops or 0.1 g of the unknown compound. Shake and allow to stand at room temperature for 3 min. Note the time for the precipitate to form or whether an iodine color develops. Heat if necessary.

1,2-Dibromo and 1,2-dichloro compounds give a precipitate of sodium bromide or chloride and also liberate free iodine.

$$\begin{array}{ccc} \underset{\underset{Br}{|}\quad \underset{Br}{|}}{RCH-CHR'} + 2NaI & \longrightarrow & \underset{\underset{I}{|}\quad \underset{I}{|}}{RCH-CHR'} + 2NaBr \end{array}$$

$$\begin{array}{ccc} \underset{\underset{I}{|}\quad \underset{I}{|}}{R-CH-CHR'} & \longrightarrow & RCH{=}CHR' + I_2 \end{array}$$

Ethylene dihalides react with sodium iodide in acetone at 25°C as shown in Table 5-2.

TABLE 5-2   Reaction Times of Ethylene Dihalides with NaI in Acetone at 25°C

Compound	Reaction time, min
1,2-Dibromoethane..............	1.5
1-Bromo-2-chloroethane	3.0
1,2-Dichloroethane..............	None (25 min at 50°C)

Picryl chloride has been reported by Blatt and Tristam[16] to give picryl iodide with potassium iodide in ethanol. Treatment with boiling acetic acid and alcoholic potassium iodide or even acetic acid in acetone at room temperature with this reagent produces trinitrobenzene and free iodine.

Iodine is liberated with this reagent, and sodium bromide is precipitated upon reaction with polybromo compounds, e.g., bromoform and s-tetrabromoethane, at 50°C. Carbon tetrabromide also reacts with this reagent at 25°C, although carbon tetrachloride is relatively unreactive.

The sulfonyl chloride gives an almost instantaneous precipitate with this reagent and evolves free iodine, probably due to excess sodium iodide reacting on the sulfonyl iodide:

$$ArSO_2Cl + NaI \longrightarrow ArSO_2I + NaCl{\downarrow}$$
$$ArSO_2I + NaI \longrightarrow ArSO_2Na + I_2$$

Aryl sulfonates show precipitates due to the formation of sodium sulfonate:

$$ArSO_2OR + NaI \longrightarrow ArSO_2ONa + RI$$

In most of these reactions, the products of the combinations can be examined by the chromatograph. It may be necessary to convert some of the products into derivatives that are more volatile or less easily decomposed by heat.

Hydrolysis. Almost all alkyl halides hydrolyze more easily than the aryl halides.

Mix 0.1 g of the halogen compound with 5 ml of alcoholic potassium hydroxide solution (5%) and reflux the mixture for 5 min. Cool the reaction and add 10 ml of distilled water. Acidify with dilute nitric acid (1 : 10) and filter if necessary. Add 2 drops of silver nitrate solution (5%). Aryl halides produce little or no precipitate.

The alkyl group is usually hydrolyzed to an alcohol and may be chromatographed (see Fig. 5-64).

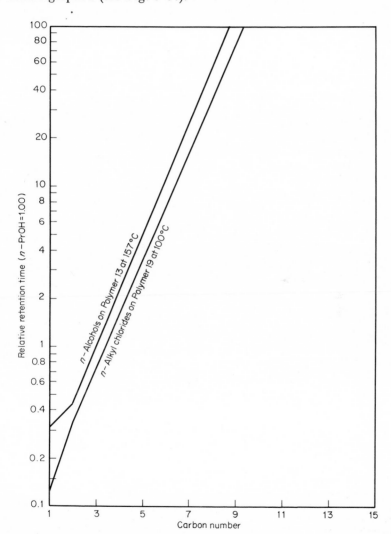

Fig. 5-64 Separation of the hydrolysis product of an alkyl halide (as an alcohol) on Poropak Q column (3 ft × ⅛ in.) at 80°C.

Mercaptan Formation. Similarly, as with silver nitrate and sodium iodide reagents, alkyl halides are much more reactive to sodium or potassium disulfide, to produce mercaptans, than the aryl halides. Some reactive aryl halides produce thiophenols.   This test is useful for alkyl compounds up to eight-carbon chain length; above this length the characteristic mercaptan odor is replaced by odors resembling those of

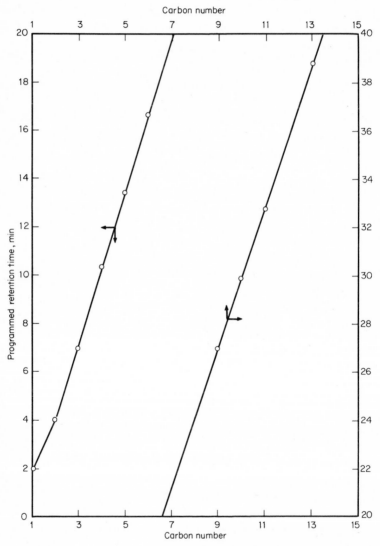

Fig. 5-65   Separation of mercaptans on Triton X-305 (20%) column (6 ft × ¼ in.) from 30 to 100°C at 10°/min.[27]

the analogous alcohol homologs. The solutions may be injected into the instrument and the resultant chromatograms compared with known similar homologs (see Fig. 5-65).

Dissolve 0.2 g of the unknown halogen compound in 2 ml of ethanol and add 2 ml of disulfide reagent (10%) made by bubbling hydrogen sulfide into sodium or potassium hydroxide solution (10%) until the pH reaches 8. **Cautiously** smell the solution. If the mercaptan odor is not detected, warm the solution and again cautiously check odor. The thiophenols are not found unless the solution is boiled and smells like mercaptans. Inject the solution into the chromatograph and observe the retention time of the emerging peak. Compare with known compounds.

Compounds with Sulfur in Major Functional Group

The presence of sulfur together with the odor, solubility, and chromatographic examination is useful in classifying the compound.

MERCAPTANS

Most low-molecular-weight mercaptans and their aromatic homologs, the thiophenols, are insoluble or only slightly soluble in water. However, they dissolve easily in sodium hydroxide solution reagent to form soluble sodium salts. Use **caution** in smelling the mercaptans or thiophenols as some are toxic in large concentrations. All have penetrating skunklike odors. Handle these compounds only under an efficient hood. The thiophenols can be detected by suitable tests for their aromatic portion, e.g., ready bromine substitution or easy nitration.

Isatin Test

A distinct green color is produced by this reagent in the presence of mercaptans. Sulfides, e.g., free hydrogen sulfide or alkyl sulfides, do not interfere, as they do not produce the green color.

To 2 ml of 1% isatin in concentrated sulfuric acid add a few drops of a dilute solution of the mercaptan in ethanol. If a mercaptan or thiophenol is present, a green color is produced. The product of the reaction cannot be examined by the chromatograph due to the presence of sulfuric acid, which causes complete decomposition in the injection port of the instrument and possible damage to the equipment.

Lead or Mercury Mercaptide Test

Nearly all mercaptans react with lead acetate or mercuric cyanide to form their corresponding lead or mercuric mercaptides. Other salts of weak acids may also be used.

$$2RSH + Pb(CH_3COO)_2 \longrightarrow Pb(SR)_2 + 2CH_3COOH$$
$$2RSH + Hg(CN)_2 \longrightarrow Hg(SR)_2 + 2HCN$$

Add several drops of the unknown mercaptans to 5 ml of saturated lead acetate solution or mercuric cyanide solution in ethanol. If mercaptans are present, a yellow precipitate is formed with the lead acetate reagent. Mercuric mercaptides produce similar precipitates but with different colors; they may vary from red to yellow to black.

Doctor Test (Lead sulfide)

In refinery operations, the doctor test is used on sour distillates to indicate the presence of mercaptans. Sodium plumbite solutions first form yellow precipitates of lead mercaptides. Free sulfur converts the lead salt to black lead sulfide and the mercaptans to alkyl disulfides:

$$Na_2PbO_2 + 2RSH \longrightarrow Pb(SR)_2\downarrow + 2NaOH$$
$$Pb(SR)_2 + S \longrightarrow PbS\downarrow + RSSR$$

To 2 ml of sodium plumbite solution add 1 drop of the unknown mercaptan and shake vigorously. A yellow precipitate of lead mercaptide forms if mercaptans are present. Add 0.05 g of flowers of sulfur (finely powdered). The color may change first to orange then gradually to black if mercaptans are present. The alkyl disulfides can be chromatographed, as shown in Fig. 5-66.

Nitroprusside Test

Mercaptans added to an alkaline solution of sodium nitroprusside give a deep wine-red color. Hydrogen sulfide gives this same reaction. Alkyl sulfides produce a color which is more reddish than blue. Thiophenols will give the reaction if ammonium hydroxide is used to alkalize the nitroprusside reagent. Aryl sulfides are negative.

To 1 drop of the unknown mercaptan in 2 ml of sodium nitroprusside solution (1%) add 3 drops of sodium hydroxide solution (10%). For thiophenols use ammonium hydroxide. If mercaptans are present, a deep wine-red color forms, which turns yellow on acidification with hydrochloric acid.

These solutions occasionally can be chromatographed for the reaction products.

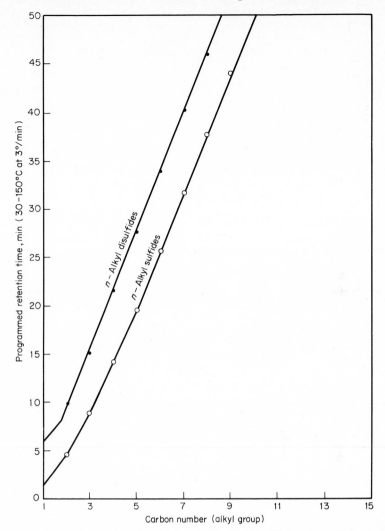

Fig. 5-66 Separation of alkyl disulfides on Triton X-305 (20%) column (6 ft $\times$ $\frac{1}{4}$ in.) from 30 to 100°C at 10°/min.[27]

SULFATES

Caution: Most of the liquid alkyl sulfates are toxic.   Avoid breathing the fumes or spilling the solutions on the skin.

Alkyl compounds of sulfuric acid can be hydrolyzed by refluxing with hydrochloric acid solution (10%) for 10 to 15 min.   The sulfate can be detected upon adding barium chloride solution (10%) by the formation of a white precipitate.   The organic portion can be detected by injecting the water layer or the oily layer (if higher alkyl groups than butyl are present) into the instrument (see Fig. 5-67).

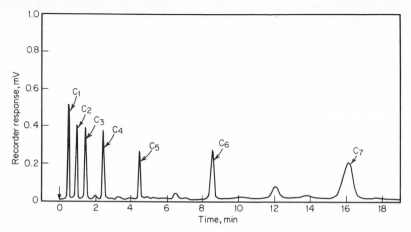

Fig. 5-67 Separation of hydrolysis product of alkyl sulfates on Poropak Q column (3 ft × ⅛ in.) at 120°C.

The salts obtained from the neutralization of sulfuric acid by organic bases can be decomposed using sodium hydroxide solution. The liberated bases can be extracted with ether. The water layer is acidified with hydrochloric acid. The sulfate is detected by the precipitate formed upon the addition of barium chloride solution (10%). The bases can be chromatographed as described under amines.

Mix 0.3 g of the unknown sulfate salt with 5 ml of sodium hydroxide solution (10%). Add a few milliliters of ether and extract the free bases into the ether layer. Acidify the water layer with acetic or hydrochloric acid and add 1 ml of barium chloride solution (10%). A white precipitate of barium sulfate forms if the salt is a sulfate.

Most organic sulfates cannot be gas-chromatographed as is since they decompose readily in the hot zone of the injection port. Usually the free base or alcohol is separated from the sulfuric acid and examined chromatographically.

SULFIDES

The organic sulfides differ from the mercaptans in odor and solubility properties, but the chemical tests of the two are quite similar. For example, in the doctor test, the first precipitate is light yellow instead of golden yellow but does not turn black (unless hydrolyzed). The precipitate does turn orange upon the addition of free sulfur. In the nitroprusside test, sulfides tend to produce red rather than wine-colored solutions, which tend to turn yellow.

The retention times and chromatographic behavior of the sulfides and mercaptans differ greatly (see Fig. 5-68).

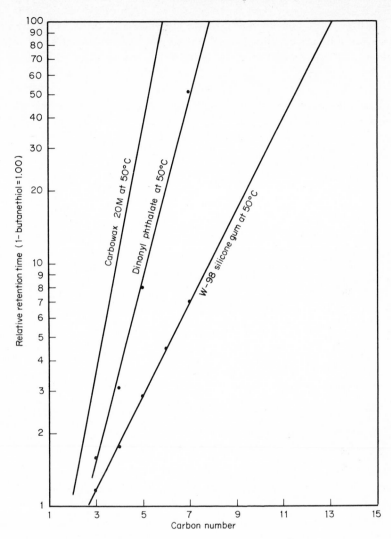

Fig. 5-68   Separation of *n*-alkyl sulfides on different columns.[27]

SULFONAMIDES

When sulfonamides are fused with solid sodium hydroxide, free ammonia or amines are liberated (if the latter unknown is an *N*-substituted sulfonamide):

$$C_6H_5SO_2NH_2 + 2NaOH \longrightarrow C_6H_5ONa + NaHSO_3 + NH_3$$
$$C_6H_5SO_2NHR + 2NaOH \longrightarrow C_6H_5ONa + NaHSO_3 + NH_2R$$

The free ammonia or amine in the distillate can be detected by its effect on red litmus paper or chromatographed (as described under amines). If the solid residue is dissolved in water and acidified, free sulfur dioxide can be detected by injection of the water layer into the instrument (under conditions shown in Fig. 5-69) and by its ability to oxidize green nickel(II) hydroxide to a grayish-black nickel(III) hydroxide. Benzidine acetate will increase the sensitivity of the test. The free phenol formed may be examined as is (see under phenols) or as derivatives.

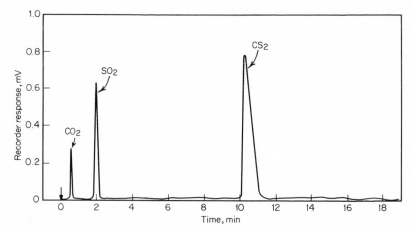

Fig. 5-69 Separation of free sulfur dioxide on Poropak Q column (3 ft × ⅛ in.) from 60 to 210°C at 10°/min.[120]

Prepare nickel(II) hydroxide by precipitation with sodium hydroxide solution and a nickel(II) salt. Wash with distilled water until the filtrates are no longer alkaline. Prepare a benzidine acetate solution by dissolving 0.05 g of benzidine in 60 ml of acetic acid (10%). Apply the nickel(II) hydroxide to filter-paper strips just before use.

Using a flameproof test tube, fuse 0.5 g of the unknown sulfonamide with 3 g of sodium hydroxide pellets. Test the gas over the fusion mass for ammonia or amines (collect in water for further tests). Cool the fusion mass and dissolve it in distilled water. Acidify the solution with hydrochloric acid and suspend a strip of filter paper treated with a paste of nickel(II) hydroxide. Heat the tube to assist the evolution of sulfur dioxide. A change from green to grayish-black indicates that oxidation has taken place. Add a drop of benzidine acetate solution to the filter-paper strip. Oxidation has taken place when the paper turns bright blue. Disregard a color that develops after a few minutes.

It is essential that both tests for sulfur dioxide and ammonia be positive, as many other sulfur-containing compounds liberate free sulfur dioxide. The few substances other than sulfonamides that can give positive tests for both can be differentiated by variations in solubilities and chromatographic differences.

The sulfonamides can be successfully separated by chromatographic examination if one is certain not to exceed the decomposition temperatures in the injection port, column, and detector. The sulfonamides can also be transformed into trimethylsilyl derivatives, which are more volatile and somewhat more stable. Several series of sulfonamides are shown in Fig. 5-70 as is and as trimethylsilyl derivatives.

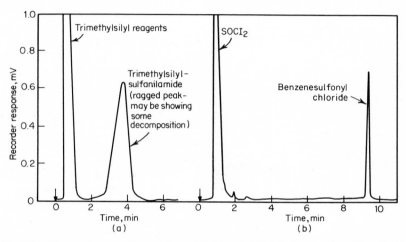

Fig. 5-70 Separation of sulfonamides as (*a*) trimethylsilyl derivatives and (*b*) sulfonyl chlorides on W-98 (10%) column (6 ft × ¼ in.) from 80 to 240°C at 10°/min.

SULFONIC ACIDS

Hydroxamate Ester Test

The sulfonic acids can be converted into sulfonhydroxamic acids by treatment initially with thionyl chloride and then with hydroxylamine hydrochloride:

$$ArSO_3H + SOCl_2 \longrightarrow ArSO_2Cl + HCl + SO_2$$
$$ArSO_2Cl + H_2NOH \cdot HCl \longrightarrow ArSO_2NHOH + 2HCl$$

The sulfonhydroxamic acids formed will react with acetaldehyde in an alkaline solution to form acethydroxamic and sulfinic acids. Both these acids react with ferric chloride.[99] The ferric hydroxamate is a wine-red soluble compound; the ferric sulfinate is an orange-red insoluble precipitate:

$$ArSO_2NHOH + CH_3CHO \longrightarrow CH_3CONHOH + ArSO_2H$$
$$3CH_3CONHOH + FeCl_3 + 3KOH \longrightarrow Fe(CH_3CONHO)_3$$
$$+ 3KCl + 3H_2O$$
$$3ArSO_2H + FeCl_3 \longrightarrow Fe(ArSO_2)_3 + 3KCl + 3H_2O$$

Sulfonic acid salts may be neutralized with hydrochloric acid and then evaporated to dryness on a steam bath before treatment with thionyl chloride.

To 0.1 g of the unknown sulfonic acid add 5 drops of thionyl chloride and place the test tube in a boiling water bath under a hood for 1 min. Cool and add 0.5 ml of a methanol solution of hydroxylamine hydrochloride (7%) and 1 drop of acetaldehyde. Alkalinize by the addition of 2 N methanolic potassium hydroxide. Heat to boiling and cool. Acidify with dilute hydrochloric acid and add 1 drop of ferric chloride solution (5%). A wine-red color with or without a brown-red precipitate is considered a positive test for sulfonic acids.

Sulfonic acids usually cannot be chromatographed successfully as is due to decomposition. However, conversion into the sulfonchloride (or sulfonfluoride[109]) allows them to pass through the instrument without any substantial decomposition (see Fig. 5-71). Other derivatives, such as the trimethylsilyl, are stable and volatile enough to pass through

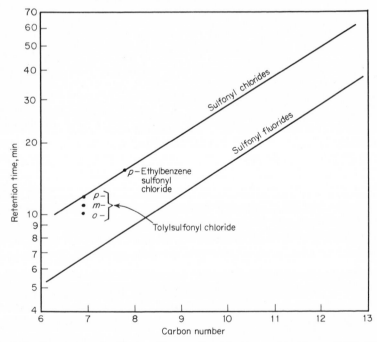

Fig. 5-71 Separation of sulfonic acids as sulfonchlorides or fluorides on 3.8% SE-30 column (4 ft $\times$ ¼ in.) at 170°C.[141]

the instrument unchanged.   If the sulfonhydroxamic acid is trimethyl-silylated, it can also be successfully chromatographed, as shown in Fig. 5-72.

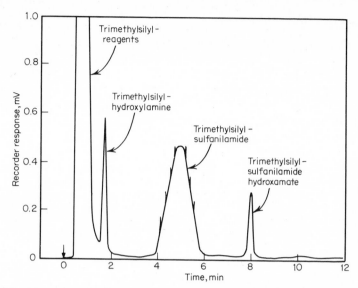

Fig. 5-72 Separation of sulfonhydroxamic acid as a trimethylsilyl derivative on W-98 (10%) column (6 ft × ¼ in.) from 60 to 220°C at 10°/min.

Fusion Test

All the sulfonic and sulfinic acids and sulfones release sulfur dioxide when fused with sodium hydroxide, as described under sulfonamides. The sulfur dioxide can be detected by its oxidation of nickel(II) hydroxide or by chromatographic examination.[141]   The sulfonic and sulfinic acids can be differentiated from the sulfones by differences in their solubilities, acidity of their solutions, and chromatographic behavior.

If sulfides are present, the odor of hydrogen sulfide will be evident, and black nickel sulfide will appear on the strip of treated filter paper, as described under sulfonamides.

All the degradation products of the sulfonic and sulfinic acids, sulfones, and other sulfur-containing compounds can be observed by chromatography, either as is after separation from the neutralized fusion mass or after conversion to suitable derivatives.

If the compound is an aryl sulfonyl chloride or a sulfonic acid which has been converted to the sulfonchloride, it can be reacted with known primary or secondary amines.   The chromatogram or the melting point of the resulting amide can be compared with known pure derivatives.

Miscellaneous Functional Tests

POLYHYDROXY COMPOUNDS

Many polyhydroxy compounds, including glycols, polyols, carbohydrates, and sugars, can be oxidized by cold periodic acid solution yielding formaldehyde, formic acid, water, and other acids or aldehydes:[47]

$$\begin{matrix} \text{RCHOH} \\ | \\ \text{RCHOH} \end{matrix} + \text{HIO}_4 \longrightarrow 2\text{RCHO} + \text{H}_2\text{O} + \text{HIO}_3$$

$$\begin{matrix} \text{RCHOH} \\ | \\ \text{RC=O} \end{matrix} + \text{HIO}_4 \longrightarrow \text{RCHO} + \text{R'COOH} + \text{HIO}_3$$

$$\begin{matrix} \text{RC=O} \\ | \\ \text{RC=O} \end{matrix} + \text{HIO}_4 \xrightarrow{\text{H}_2\text{O}} 2\text{RCOOH} + \text{HIO}_3$$

The aldehyde can be detected by using Schiff's reagent. Citric acid is negative to this procedure, but tartaric acid gives a positive reaction.

Place 2 ml of the periodic acid reagent (0.5 g paraperiodic acid in 100 ml of distilled water) in a test tube and add 1 drop of concentrated nitric acid. Mix thoroughly and add 1 drop of liquid or an equivalent crystal of the unknown compound to be tested. Shake vigorously for 10 to 20 sec and then add 1 to 2 drops of aqueous silver nitrate solution (5%). An instant white precipitate indicates formation of iodate by oxidation of the organic substance and reduction of the periodic acid. If no precipitate forms or a brown precipitate forms which redissolves on shaking, the test is considered negative.

Iodic acid formed in this reaction can oxidize the lower homologous-series alcohols, aldehydes, methyl ketones, phenols, and aniline derivatives.[194] The iodic acid is further reduced to hydriodic acid; the hydriodic acid and iodic acid interact to produce free iodine. Thus, the brown color of free iodine may result.

The iodic acid and excess reagent can be destroyed by adding several drops of saturated sulfurous acid solution. A portion of the solution may be treated with a few drops of Schiff's reagent. A red-blue color will develop within 30 min if aldehydes are present. Polysaccharides give this test if the mixture is heated to boiling before adding the Schiff's reagent. Schiff's reagent is a selective oxidant on 1,2-glycols, α-hydroxy aldehydes, α-hydroxy ketones, 1,2-diketones, and α-hydroxy acids, the rate decreasing in the order given.[1]

Although olefins, secondary alcohols, 1,3-glycols, ketones, and alde-hydes are not oxidized by periodic acid under these conditions, higher temperatures will bring about oxidation. This procedure is best suited for water-soluble compounds. Using dioxane solutions is better for insoluble compounds.

The epoxy group can also be detected by this reagent, providing that it is hydrolyzed to the glycol. In examining for the epoxy group, dissolve 2 drops of the unknown (or the equivalent crystals) in 2 ml of glacial acetic acid and add the mixture to 2 ml of periodic acid reagent solution (0.5%) that has been acidified with 2 drops of concentrated nitric acid. Shake the mixture thoroughly and then dropwise add 2 to 3 drops of silver nitrate solution (5%). A white precipitate is considered a positive test.

The oxidation products can be examined by the chromatograph as is or further derivatized as methyl ester (if acids) or as trimethylsilyl compounds and other volatile substances (see under aldehydes, acids, etc.).

TESTS FOR UNSATURATION AND SUBSTITUTION

Bromine in Carbon Tetrachloride

This procedure is useful for detecting unsaturated bonds or substitution in aromatic compounds. To 0.1 g (or 0.2 ml of liquid) of the unknown substance add 2 ml of carbon tetrachloride. Dropwise add a 5% solution of bromine in carbon tetrachloride with shaking until the bromine color persists. A positive test for unsaturation is the de-colorization of the bromine without the evolution of hydrogen bromide (white fumes). A positive test for substitution is decolorization of the bromine with the evolution of hydrogen bromide (white fumes).

Bromine in Water

Using a polar solvent (water) instead of a nonpolar solvent (carbon tetrachloride) greatly increases the rate of bromination because bromine ionizes more readily in water.[195] However, the evolution of hydrogen bromide is not readily detected by visual observation. It can be noted by testing the aqueous solution for increase in acidity or observing the hydrogen bromide peak on the chromatograph.

Besides the usual addition and substitution reactions, bromine will oxidize a number of substances, including mercaptans, to disulfides:

$$2RSH + Br_2 \longrightarrow RSSR + 2HBr$$

OXIDATION REAGENTS

A large number of oxidants are used in organic chemistry.

Copper(II) Ion

The various copper(II) reagents used in oxidizing organic compounds differ only in regard to their pH and the salts used to prevent copper-ion precipitation.

Benedict's Reagent. This reagent[12] is less alkaline than Fehling's solution and can be stored on the shelf as one solution. The copper(I) oxide from both reagents is identical, the color varying from bluish-green (for very fine particles) to red (for the largest particles). Yellow-orange is the most common color.

Add 0.1 to 0.2 g of the substance to be oxidized to 5 ml of water and 5 ml of Benedict's reagent. Immerse the tube in boiling water so that the level in the tube is below the level of the water and keep it there for 5 min. An easily oxidized substance will yield a yellow, orange, or red precipitate. Neutralize the supernatant liquid and examine the oxidized material in the chromatograph (see acids, aldehydes, etc.).

Fehling's Reagent.[91] Just before use, mix 2.5 ml of Fehling's solution A with 2.5 ml of solution B and shake until clear. Add 5 ml of distilled water and 0.1 to 0.2 g of the material to be oxidized. Boil gently for 2 min. A precipitate of copper(I) oxide is confirmation of the presence of an easily oxidizable substance. Examine the neutralized supernatant for the oxidation products by the instrument, keeping in mind the presence of tartaric acid in the solution.

Silver Ion

Tollens' Reagent. This reagent is an alkaline solution of silver complexed with ammonia ions. The silver ion is reduced to free silver by nearly all aldehydes, readily oxidizable sugars, aminophenols, polyhydroxy phenols, hydroxylamine, and other reducing compounds.[132]

Boil a clean test tube for 1 min with 5 ml of sodium hydroxide solution (10%). Discard the solution and add to the thoroughly clean tube 2 ml of silver nitrate solution (5%). Add a drop of sodium hydroxide solution (10%). Dropwise, add a dilute solution of ammonium hydroxide until the precipitate just dissolves. Add 0.1 g of the unknown compound and mix well. *Do not heat the solution.* Allow it to stand for 10 min. A silver mirror or a precipitate of silver indicates that the substance has been oxidized. Neutralize a portion of the supernatant liquid with hydrochloric acid, and filter. Inject the clear filtrate into the chromatograph and examine for aldehydes, ketones, acids, and other oxidized products. Discard the balance of the solution and flush with copious amounts of water, as the dry residue may explode.

Iodic Acid

Iodic acid[194] is a fairly selective oxidizing agent. Simple alcohols up to C_7 (exception is methanol), aldehydes, methyl ketones, phenols, and derivatives of aniline are oxidized by this reagent under the conditions of the test. Polyhydroxy alcohols (except 1,2-propanediol and 1,3-propanediol), acids, and sugars (except fructose and sucrose) are not oxidized by this reagent:

$$3C_3H_7OH + HIO_3 \longrightarrow 3CH_3CH_2CHO + HI + 3H_2O$$
$$3CH_3CH_2CHO + HIO_3 \longrightarrow 3CH_3CH_2COOH + HI$$
$$5HI + HIO_3 \longrightarrow 3I_2 + 3H_2O$$

To prepare the reagent, slowly add 2.5 ml of concentrated sulfuric acid to 8 ml of distilled water. Cool to room temperature and add 0.1 g of potassium iodate. Add 0.05 g of the unknown compound to this reagent and immerse the tube in a boiling water bath for 1 hr. A brown color of free iodine indicates that oxidation has taken place. Examine the neutralized solution for oxidized products and free iodine by the chromatograph.

Potassium Permanganate Solution

Potassium permanganate solution is decolorized by substances having ethylenic or acetylenic groups. The unsaturated group is first oxidized to the glycol, and further oxidation cleaves the compound at the unsaturation linkage:

$$RCH{=}CHR' \xrightarrow{(O)} \begin{array}{cc} RCH-CHR' \\ | \quad | \\ OH \quad OH \end{array} \longrightarrow \begin{array}{cc} R-C{=}O + O{=}C-R' \\ | \qquad\qquad | \\ OH \qquad\quad OH \end{array}$$

Acetylenic compounds are cleaved by the permanganate oxidation to yield acids:

$$R-C{\equiv}C-R' + H_2O + 3(O) \longrightarrow RCOOH + R'COOH$$

As the reaction proceeds, potassium hydroxide is liberated from the potassium permanganate and the reaction becomes more alkaline. Use of sodium carbonate solution changes the nature of the reaction. Acetone (normally negative) gives a positive test in the presence of sodium carbonate. This reagent is used to split a molecule at a non-cyclic double bond or oxidize an alkyl side chain to a carboxylic acid group.

Use of zinc permanganate in a neutral medium will provide a nearly neutral oxidation atmosphere, as the zinc hydroxide formed is only slightly soluble. Oxidation with potassium permanganate in the presence of magnesium sulfate, which precipitates magnesium hydroxide, essentially keeps the solution neutral.

Alcohols also decolorize potassium permanganate solutions but do not decolorize bromine solutions. Pure alcohols do not decolorize potassium permanganate. Only alcohols containing impurities that can oxidize readily will decolorize the permanganate. This reagent in aqueous acetic acid is used to differentiate the primary from the secondary and tertiary alcohols.[152] The primary and secondary alcohols react, but the tertiary alcohols do not.

Organic sulfur-containing compounds in which the sulfur is in the lower oxidation states can be oxidized by potassium permanganate:

$$2RSH + (O) \longrightarrow RSSR + H_2O$$

$$RSR + (O) \longrightarrow R_2SO + (O) \longrightarrow RSO_2R$$

$$R_2C(SR')_2 + 4(O) \longrightarrow R_2C(SO_2R')_2$$

If the solution is acidic, sulfur compounds are more readily oxidized by potassium permanganate. Dissolve about 0.1 g of the sulfur-containing compound in 2 ml of glacial acetic acid and dropwise add the solution of potassium permanganate. Decolorization indicates that an oxidizable sulfur may be present in the molecule. Sulfur at higher oxidation states, e.g., in sulfones, alkyl sulfates, or sulfonic acids (unsubstituted), is not oxidized further by this reagent. Certain substituted sulfonates, e.g., the aldehyde or ketone bisulfite addition compound and phenolic sulfonic acids, do decolorize permanganate solutions.

The oxidation products can be separated from the reaction mixture and examined by the chromatograph (as is if volatile enough, or converted into derivatives which are more volatile). For example, the oxidation products of phenol can be chromatographed on a variety of columns. The acids can be observed on nonpolar columns, such as Poropak Q (see Fig. 5-73), or as the methyl derivatives on the Apiezon L column. The carbon dioxide can be collected and chromatographed on a silica gel column at 25°C.[118]

Potassium Dichromate and Sulfuric Acid

This reagent is generally used when oxidizing alcohols. For example, when ethanol is added to an acidic solution of potassium dichromate and warmed slightly, the ethanol is immediately oxidized to acetaldehyde

$$CH_3CH_2OH \xrightarrow[K_2Cr_2O_7]{(O)} CH_3CHO$$

Further treatment with more acidic potassium dichromate converts the acetaldehyde into acetic acid.

The aldehydes may be steam-distilled out of the reaction mixture or extracted with ether and chromatographed, as shown in Fig. 5-12.

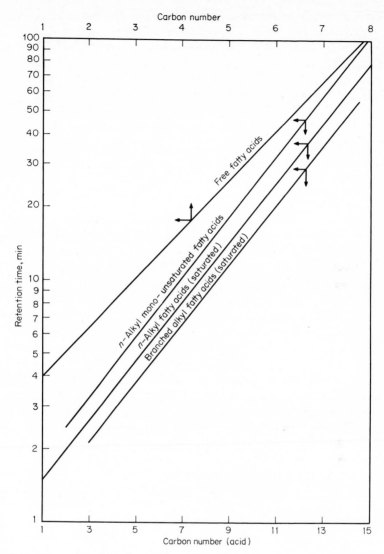

Fig. 5-73 Separation of free fatty acids on Poropak Q column (3 ft × ⅛ in.) from 60 to 220°C at 10°/min and separation of fatty acids as methyl esters on a silicone polyester column, Hi-Eff 8BP (3%) plus W-52 phenyl silicone (5%) (3 ft × ¼ in.) from 60 to 220°C at 10°/min.

Picric Acid

An alkaline solution of picric acid will be reduced from yellow picrate to an amber-red picramate ion with consequent oxidation of any oxidizable organic groups.

Place 2 ml of a saturated aqueous solution of picric acid in a test tube in a boiling water bath. Add 1 ml of sodium hydroxide solution (5%) and 0.1 to 0.05 g of the unknown substance. Keep the solution in the boiling water bath for 30 min. A positive test is a change in color from the yellow picrate to the red color of the picramate ion. This reaction involves the reduction of one nitro group to an amino group.

Hydrogen Peroxide

The reagent is available commercially in concentrations from 30 to 100 percent. The 90% solution is highly purified and does not require a stabilizer. **Caution**: Reacting any unknown material with hydrogen peroxide should be performed behind a safety shield using small quantities of reactant and reagent.[68]

Acidic. This reagent oxidizes olefinic groups to epoxides. The epoxide ring is then cleaved in the presence of formic acid, producing a diol monoformate. The free diol is separated by hydrolysis using a solution of sodium hydroxide. The diol, if volatile enough, can be examined as is on the chromatograph or converted into an acetate or other volatile derivative.

Hydrogen peroxide in strongly acidic medium oxidizes diisobutylene to neopentyl alcohol and acetone.[95] Swern[171] determined the rate of oxidation of a variety of unsaturated compounds by hydrogen peroxide in acetic acid. Certain o-quinones can be oxidized by this reagent in warm acetic acid with subsequent cleavage of the carbon-carbon bond.[97] Aromatic hydrocarbons have also been oxidized to quinones by hydrogen peroxide in acetic acid.[3] Aryl amines, nitroso compounds, and azobenzenes can be oxidized by hydrogen peroxide in acetic acid, converting them to nitroso-, nitro-, and azoxybenzene compounds, respectively.[98,111,135] Most of these products can be examined on the chromatograph as is or further derivatized.

Hydrogen peroxide in the presence of boron trifluoride etherate cléaves aliphatic ketones to form esters.[121] Pyridine N-oxides can be made by oxidation of pyridine with hydrogen peroxide in acetic acid.[173]

Basic. In basic hydrogen peroxide, quinones can be oxidized to epoxides.[57] Hydroquinone can be oxidized[103] to a dihydroxybenzoquinone in strong sodium hydroxide solution.

Hydrogen peroxide will oxidize salicylaldehyde to catechol.[39] Alkaline hydrogen peroxide will cleave α-keto acids and α-ketones.[165] Most of these products can be observed chromatographically as is or further derivatized as esters, etc.

Base-catalyzed hydrogen peroxide will convert the nitriles to the corresponding amides.[138] These compounds can be observed on the chromatograph or converted to the trimethylsilyl derivatives.

Neutral. Neutral hydrogen peroxide will oxidize amines to amine oxides.[36] These products are usually too unstable to be chromatographed successfully.

Thioethers can be oxidized to sulfoxides by neutral hydrogen peroxide in acetone (the sulfones are produced with this reagent in 50% acetic acid),[69] and sulfides are oxidized to disulfides. Triphenylarsine can be oxidized to the oxide by neutral hydrogen peroxide. Many of these products can be chromatographed satisfactorily.

With Iron Catalyst. Hydrogen peroxide catalyzed with ferrous sulfate converts *t*-butanol into 2,5-dimethylhexane-2,5-diol.[101] Aniline in the presence of ferrous ion, ethylenediaminetetraacetic acid, and ascorbic acid is oxidized to *p*-hydroxyaniline.[181]

With Vanadium Catalyst. Vanadium pentoxide in the presence of hydrogen peroxide oxidizes cyclohexane to cyclohexanone and Δ-cyclohexene-1-ol.[179]

With Tungsten Catalyst. A primary amine containing the $-CH_2NH_2$ group is oxidized to the aldoxime by hydrogen peroxide (35%) in the presence of the catalyst sodium tungstate.[106]

Sulfides can be oxidized to sulfones by hydrogen peroxide in acetic acid catalyzed by sodium tungstate.[158]

With Osmium Tetroxide Catalyst. Olefins can be oxidized to *cis*-diols by hydrogen peroxide in *t*-butanol in the presence of catalytic amounts of osmium tetroxide.[129]

With Selenium Dioxide Catalyst. Olefins may be similarly oxidized with hydrogen peroxide in the presence of selenium dioxide in *t*-butanol at 0°C (65 hr).[168]

Acrolein is oxidized by hydrogen peroxide in the presence of selenium dioxide in t-butanol to acrylic acid.[164] This and other oxidation products can be chromatographed as is if they are stable enough to pass through the instrument unchanged; or they can be converted to esters or trimethylsilyl derivatives.[143]

Miscellaneous Oxidizing Agents

Ceric Nitrate. Ceric ammonium nitrate oxidizes alcohols with a change of color of the reagent from yellow to red. Phenols give a brown to green-brown precipitate in water solutions; a deep red to brown color is formed in dioxane solutions.

The reaction is adaptable to alcohols and phenols with up to 10 carbons in their chains. The larger molecules are usually too insoluble and the hydroxyl group too diluted by the hydrocarbon chain to give a good color change. Hydroxy acids or hydroxy aldehydes and most substances containing alcoholic hydroxy groups of 10 carbons or less give a positive test. Amino alcohols usually precipitate the ceric ion unless sufficiently acidified.

Acids, aldehydes, alkyl halides, esters, ketones, and other compounds containing carbon, hydrogen, oxygen, and halogen do not normally interfere. Amines as hydrochlorides, aromatic amines, and similar compounds are capable of giving color groups or precipitates with this reagent.[44]

Benzyl alcohols are oxidized to benzaldehydes by ceric reagent in aqueous acetic acid (50%) upon warming. Toluenes are also oxidized to benzaldehydes by this reagent in a solution of aqueous acetic acid (50%). With glacial acetic acid, the toluenes are converted to benzyl acetates by the ceric reagent.[178]

To prepare the reagent dissolve 40 g of ceric ammonium nitrate, $Ce(NH_4)_2(NO_3)_6$, in 100 ml of warm 2 N nitric acid. Mix 0.5 ml of the reagent with 3 ml of distilled water. Add 4 to 5 drops of the unknown substance (or 4 to 5 drops of an aqueous dioxane solution of the solid substance). Shake the solution vigorously. A change from yellow to red is a positive indication that oxidation has taken place.

The oxidant can be observed by injection of the neutralized solution into the instrument. Most aldehydes are sufficiently volatile to be observed. If the compound is not volatile enough, convert it into a more volatile derivative such as the trimethylsilyl. The retention-time differences between the aldehydes and their corresponding alcohols can be observed by comparing the substances before and after oxidation by this reagent (Fig. 5-74).

Oxygen and Ozone. Oxygen and ozone have been used to oxidize

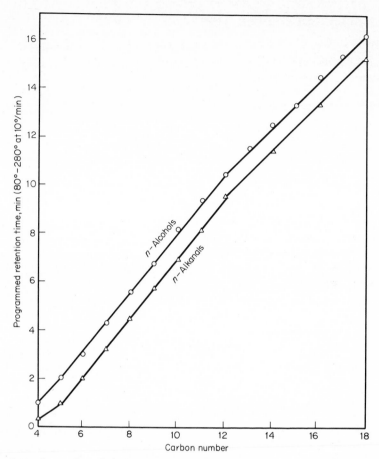

Fig. 5-74 Separation of alcohols and aldehydes (before and after oxidation by ceric nitrate) on W-98 (10%) column (6 ft × ¼ in.) from 60 to 220°C at 10°/min.

organic compounds. Unsaturated groups have been oxidized by ozone with subsequent cleavage of the carbon-carbon bond.[188]

Atmospheric oxygen or pure oxygen readily oxidizes such oxidizable substances as certain aldehydes (benzaldehyde) or alcohols. However, pure oxygen should *never* be mixed with highly combustible compounds except in an oxygen bomb. Oxidations by pure oxygen are difficult to control.

Many other substances are capable of oxidizing organic compounds, but the most important ones have already been cited. With a gas chromatograph the analyst can follow the course of the oxidations very closely, thus preventing the reaction from proceeding too far.

MISCELLANEOUS COLOR TESTS

A large variety of color tests has been published,[48] but mention will be made only of those which can be directly applied to the identification of a group or class of compounds.

Cold Concentrated Sulfuric Acid

This reagent produces a great variety of colors, especially with well-substituted or mixed cyclics. Unfortunately the colors are not specific for one class of compounds, but they may be of assistance in identifying a certain compound. To use this reagent, mix a few drops of the unknown substance with 1 ml of cold concentrated sulfuric acid in a test tube and shake. After a few minutes observe the colors produced. On warming, compounds that do not give colors with the cold acid may develop characteristic colors useful for identity tests. A number of compounds char with this acid and produce tan, brown, or black colors, i.e., cyclics and polyhydroxy compounds. Iodine compounds turn the hot acid a purple color.

None of these reaction products can be used directly in the chromatograph because of additional decomposition in the hot zone of the injection port. However, if the acid is carefully neutralized and the reacted organic compound separated from the salts, it is possible to examine the product or further convert it into a stable derivative.

Fuming Sulfuric Acid

Caution: Use only on compounds which are insoluble in cold concentrated sulfuric acid, e.g., the paraffins, cycloparaffins and their halogen derivatives, and simple unsubstituted aromatic compounds.

Place 2 ml of fuming sulfuric acid (20%) in a clean, dry test tube and add 1 ml of the unknown compound, stopper, and shake vigorously. Allow to stand a few minutes and observe any colors that may have developed or the development of heat. If a compound has reacted, dilute it carefully with water, cool, and neutralize with sodium hydroxide solution (10%). Separate the reacted compound and examine as is or convert it into a more volatile derivative.

1,2-Dihalogen compounds turn dark and release some free halogen in this reagent. Other compounds give different reactions with fuming sulfuric acid. For a more thorough discussion of the reagent, consult the literature.[100,163]

Ferric Chloride Reagent

Ferric chloride solution gives reddish colors with phenols and hydroxamate esters. West and Brode[189] have published the color reactions of this reagent with 75 compounds. Some investigators have allowed the effluent from the chromatograph (thought to contain phenolic compounds) to bubble through small quantities of this reagent. Peaks containing phenolic or hydroxamate esters give characteristic colors in the reagent.

Sodium Pentacyanoamine Ferroate

This reagent gives a great variety of colors with nitro and nitroso compounds, hydrazines, thioketones, α- and β-unsaturated aldehydes, and aromatic aldehydes.[49,50] Compounds suspected of being any of the above may be bubbled through this solution from the outlet of the chromatograph. The colors developed can be compared with the reactions of this reagent with known compounds.

Fluorescein Reagent

This reagent gives colors with amines, amides, nitriles, and pyrrole compounds.[51] Similarly, the reagent can be used on the effluent side of the chromatograph to assist in identifying any given peak or peaks.

1,2-Naphthoquinone 4-Sulfonate

Reactive methylene or amino groups in compounds give distinct color reactions with the reagent. Tertiary-ring basic compounds can be detected[51] by combining this reagent with methyl iodide. Similarly, this reagent is useful as a chromatographic effluent detector by bubbling the volatile fraction through the solution. The colors developed can be compared with known substances.

o-Dianisidine

Aldehydes can be detected by this reagent.[184] The gas-chromato-graphic effluents suspected of being aldehydes can be bubbled into this reagent and thus characterized by comparison with known aldehydes.

Potassium Cyanide

The m-dinitro compounds react with this reagent to give distinctive colors that help differentiate them from the o- or p-compounds.[12] This reagent is very useful in distinguishing the meta from the other isomers as they emerge from the instrument.

Reagents for Primary Aromatic Amines

Glutaconic aldehyde[79] and sodium pentacyanoaquoferriate[4] produce characteristic colors with primary aromatic amines. They can be used as effluent detector reagents.

REDUCING AGENTS

Ferrous Hydroxide

This reagent reduces nitro compounds to amines.   To 1 ml of ferrous sulfate reagent (25 g of ferrous ammonium sulfate and 2 ml of concentrated sulfuric acid in 500 ml of boiled distilled water to which an iron nail has been added) add 0.7 ml of alcoholic potassium hydroxide (15%) (30 g in 30 ml of distilled water to which is added 200 ml of ethanol).   Attach a tube and bubble illuminating gas through the solution for a few minutes to deoxygenate.   Add 10 mg of the unknown sample, turn off the gas, stopper, and shake vigorously.   A red-brown precipitate indicates that reduction has taken place.   A greenish precipitate is considered a negative test.   Partial oxidation may cause the precipitate to darken.

Nitro and nitroso compounds give a positive test as well as quinones, hydroxylamines, alkyl nitrates, and alkyl nitrites.   Highly colored compounds cannot be tested successfully as the color cannot be differentiated from the color change of the reagent.

If a positive test is indicated, inject some of the neutralized supernatant liquid into the instrument.   If the substance is a nitro compound, its retention time after reduction will differ from that of the parent compound (see Fig. 5-75).

All nitro compounds give a positive test within 30 sec, but the speed with which the compound is reduced depends upon its solubility. 1-Nitronaphthalene must be shaken for 30 sec, while nitrobenzene gives the test almost immediately (slightly soluble in water and 0.19 g/100 g of water, respectively).[90]

Zinc and Ammonium Chloride

This reagent is a milder reducing agent: nitro compounds are converted to hydrazines, hydroxylamines, or aminophenols.   The reduced substances can be detected by Tollens' reagent (reduction of silver) or examined by the chromatograph as is or derivatized.

Dissolve 0.5 ml or 0.5 g of the nitro compound in 10 ml of ethanol (50%).   Add 0.5 g of ammonium chloride and 0.5 g of zinc dust. Shake and heat to boiling.   Allow to stand for a few minutes, filter, and test the filtrate by means of Tollens' reagent.   A silver mirror is an indication of reducing substances.

When the original compound before reduction reduces Tollens' reagent, this reagent cannot be used.   Check the retention time of the compound before and after reduction.   If the reduced compound decomposes in the injection port or on the column of the instrument, convert it into a more stable derivative.

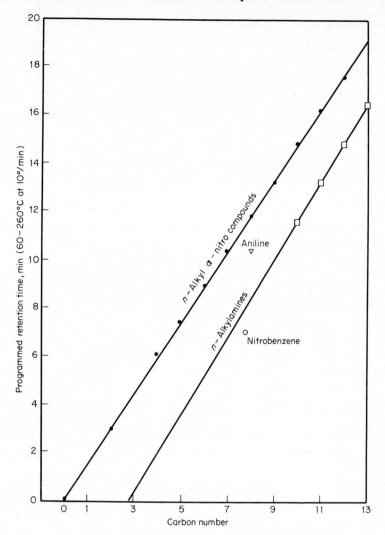

Fig. 5-75 Comparison of reduction products of nitro compounds reduced with ferrous hydroxide and separated on Chromosorb 103 column (3 ft × ⅛ in.) from 60 to 220°C at 10°/min.

Zinc or Iron and Hydrochloric or Other Acids

Zinc and hydrochloric acid reduce certain hydroxy compounds to alkanes,[37] and zinc dust and sulfuric acid reduce benzenesulfonyl chloride to thiophenol.[2]

Iron and hydrochloric acid reduce nitro compounds to amines.[46]

Zinc dust and acetic acid reduce nitroso compounds to hydrazines.[89] This reagent reduces certain double bonds to saturated bonds;[196] it also reduces certain keto groups.[122]

Zinc and Miscellaneous Media

Zinc dust in *neutral* medium suspended in an aqueous solution with
p-toluenesulfonyl chloride reduces the compound to p-toluenesulfinate.[191]
The zinc dust also reduces nitro compounds to amines[112] and certain
keto steroids to hydroxy compounds.[176]

Zinc dust in *basic* medium (in dilute alkali) reduces anthracene.[59]
Benzophenone is reduced to azobenzene;[13] o-nitroaniline can be reduced
to o-phenylenediamine;[127] and ketoximes can be reduced to primary
amines.[102]

Zinc amalgam reduces aryl alkyl ketones to the corresponding hydrocarbons.[34] Aldehydes are reduced to the corresponding hydrocarbons;[159] naphthalenesulfonyl chlorides can be reduced to the
corresponding thiols.[70]

Tin and Hydrochloric Acid

Tin and hydrochloric acid reduce the hydroxyl groups to the hydrocarbon[30] and nitro[54] and azo groups[8] to amines.

Amalgamated tin and hydrochloric acid reduce conjugated enediones
to saturated diketones.[156] This reagent is useful in a variety of re-
ductions.

Stannous chloride can selectively reduce various nitro groups on a
variety of aromatic structures.[94]

Hydrogen

Hydrogen gas can be used to reduce easily reducible groups by simply
bubbling it through the solution. Groups that are more difficult to
reduce need higher pressures and catalysts such as nickel[71] and platinum.
The catalytically activated hydrogen will reduce ethylene and acetylenic
groups; it reduces aldehydes, nitro groups, nitriles, halides, and tosylates
to hydrocarbons; it also desulfurizes compounds, etc.

Various investigators[72] have prepared precolumns of nickel or
platinum catalysts. The sample is injected into the precolumn in a
carrier stream of hydrogen. Subsequently, the substances are reduced
and swept into the analyzer column of the instrument. The reduced
compounds are detected on the effluent side of the column, as shown in
Fig. 5-76.

Miscellaneous

Lithium aluminum hydride is a very efficient reducing agent. Aldehydes,
ketones, carboxylic acids, and epoxides are reduced to alcohols; halides
to hydrocarbons; amides to amines, etc.[68] Lithium aluminum hydride
also reacts with any active hydrogen, halide, or other reactive groups.

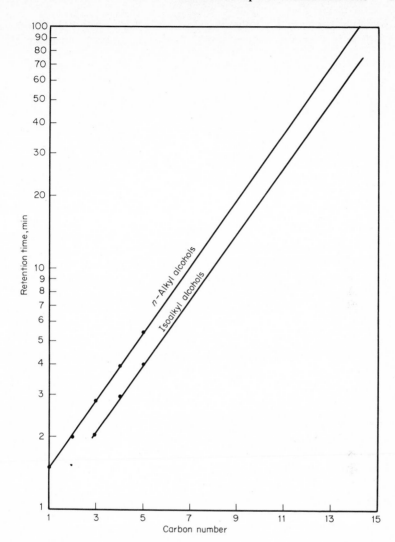

Fig. 5-76   Reduction of aldehydes to alcohols by passage over a heated precolumn of nickel (10%) (3 × ¼ in.) plus separation on Poropak Q column (3 ft × ¼ in.) at 125°C.

Lithium borohydride reagent is more reactive than sodium borohydride but less reactive than lithium aluminum hydride.   It reduces aldehydes, ketones, acid chlorides, oxides, esters, and lactones but not carboxylic acids, nitriles, or nitro compounds.[66]

Aluminum amalgam (being neutral) reduces alkali-sensitive compounds such as diethyl oxaloacetate to diethyl malate;[20] aryl alkyl ketones to pinacols;[198] and unsaturated hydrocarbons to saturated hydrocarbons.[136]

Aluminum hydride is similar in action to lithium aluminum hydride but will reduce α- and β-unsaturated aldehydes to allylic alcohols:[174]

$$C_6H_5CH{=}CHCHO \xrightarrow{\text{AlH}_3} C_6H_5CH{=}CHCH_2OH$$

The aluminum hydride will also selectively reduce the carbonyl groups of dienones.[105]

Aluminum isopropoxide selectively reduces crotonaldehyde to crotyl alcohol[73] and diaryl ketones to diaryl hydrocarbons.[192]

Lithium metal in ammonia reduces aromatic hydrocarbons to cyclic aliphatics in a stepwise fashion.[68]

Cerous hydroxide is used for reducing peroxides in ether, dioxane, or tetrahydrofuran. The wet powder turns reddish brown within 1 to 2 min, and reaction is complete in 15 min (test is negative with acidified potassium iodide–starch paper).[15]

Chromous acetate and chloride are useful for reduction of α-bromo-ketones and bromohydrins and cleavage of 2,4-dinitrophenylhydrazones.[147]

Copper powder is utilized in reductive decarboxylation and dehydration.[74]

Diborane will reduce acetone or other simple carbonyl compounds,[28] and aldehydes, epoxides, carboxylic acids, nitriles, azo compounds, and t-amides.[22]   Carbonyl chloride, nitro groups, $-SO_2-$ groups, R'X, ArX, ethers, and phenols are not reduced. Diborane reduces primary, secondary, and tertiary amides to the corresponding amines.[23]

Diisobutyl aluminum hydride is useful for the reduction of acetylenes,[21] quinoline,[193] and γ-lactones.[137]   This reagent is used in reductive hydrolysis of aryl nitriles to aryl aldehydes,[157] reduction of carboxylic acids to alcohols,[180] and reductive cleavage of α- and β-unsaturated ethers.[130]

Formaldehyde has been used to reduce other aldehydes and ketones to alcohols.[145]

Formic acid has been used to reduce triphenylcarbinol to triphenylmethane.[75]

Hydrazine will reduce keto acids to their alkyl acids.[108]   It also reduces alcohol groups on pyridoxine hydrochloride to alkyl hydrocarbon groups.[175]   In the presence of palladium, hydrazine will reduce nitro groups.[172]   In the presence of nickel, it will reduce α- and β-unsaturated acids.[144]

Hydriodic acid will reduce azlactones[67] and reduces α-diketones and α-ketals to saturated ketones.[83]

Hydroxylamine hydrochloride will reduce an aldehyde (as an oxime) to a nitrile by heating in acetic anhydride.[151]

Nickel-aluminum alloy (*Raney nickel*) in caustic soda will reduce benzyl alcohol to toluene. Acetophenone is reduced to ethylbenzene; 2-methylcyclohexanone to 2-methylcyclohexanol and oximes and nitriles to amines.[99] Ethylenic and acetylenic groups are hydrogenated in the presence of hydrogen (especially under pressure).[56] Aldehydes are reduced to alcohols and nitro and nitrile compounds to amines.[81]

Palladium and platinum hydrogenation catalysts in the presence of hydrogen will hydrogenate most systems.[60,61]

Phenylhydrazine will reduce azobenzene to hydrazobenzene at 125 to 130°C;[183] aromatic nitro compounds are reduced to arylamines.[139]

Phosphine, a highly toxic flammable gas, will reduce aromatic nitro compounds to azoxy derivatives[24] and azo compounds.[11]

Raney cobalt catalysts are used to hydrogenate simple oximes but are not as efficient as Raney nickel.[148]

Rhenium catalysts can reduce carboxylic acids to primary alcohols. Aromatic, olefinic, carbonyl, and nitro groups can be reduced by these catalysts, which, however, are less active than nickel or platinum although selective hydrogenation is possible.[19]

Rhodium catalysts (on alumina) are useful in hydrogenation of phenols to saturated cyclic alcohols.[26] Benzyl alcohols can also be hydrogenated;[167] vinylic, allylic, oxime, and cyanide groups can be hydrogenated, as well as ketals, heterocyclics, and aryl amines.[62]

Ruthenium catalysts will hydrogenate carbonyl groups, and reduce aromatic rings, olefinic and acetylenic groups, and aryl nitro groups to diaryl hydrazo compounds.[63]

Sodium amalgam reduces ketones to alcohols,[96] α- and β-unsaturated carbonyl compounds,[77] aldonolactones to aldoses,[78] and oximes to amines.[93]

Sodium hydrazide, an explosive compound, is used only in suspension in ether, benzene, or mixtures in the absence of oxygen. It reduces many unsaturated and aromatic hydrocarbons and reductively cleaves olefins.

Sodium hydride reduces ketones which have no α hydrogens.[170]

Other Reducing Agents

Sodium hydrosulfite reduces quinones,[64] and reduces azo,[55] nitro,[76] and nitroso compounds to amines;[160] it also reduces N-nitroso compounds.[140] *Sodium hypophosphite* will reduce nitriles to aldehydes in the presence of Raney nickel in aqueous acetic pyridine.[5] *Stannous chloride* will reduce nitro aryls to aryl amines;[142] it reduces quinones,[45] triphenylcarbinol,[59] and iodohydrins[38] and promotes hydrogenation.[155] *Triisobutyl aluminum* reduces ketones[122] and unsaturated lactones.[131] *Tri(n-butyl)tin hydride* reduces organic halides to hydrocarbons,[154] acid chlorides to aldehydes,[113] and carbonyl to aldehydes and ketones[114] and promotes olefinic and acetylenic unsaturation.[182] *Trimethylamine borane* reduces ketones[104] and Schiff bases.[14] *Zinc-copper couple* is used to reduce phthalimide to phthalide.[82] *Zinc hydrosulfite* is a powerful reducing agent at pH 3 to 7.5 but is most effective at pH 4 to 5 ,where the hydrosulfite is rapidly decomposed.

MISCELLANEOUS TESTS

Hydrolysis

Nitriles, amides, substituted amides, hydrazones, semicarbazones, and esters can be hydrolyzed and the resulting products identified. For example, an ester can be hydrolyzed to an alcohol and an acid. The physical and chemical properties of the free alcohols and acids, as well as those of their derivatives, and the chromatograms can be compared with those of the original ester, known esters, and their known hydrolysis products. Ethers can also be hydrolyzed with strong acids. Almost all these compounds must be hydrolyzed before derivatives can be prepared for use in establishing their identity.

Iodoform Test

Compounds which contain the CH_3CO- group or which can easily be oxidized to contain it can be converted to iodoform by an alkaline solution of sodium hypoiodite.[80] The compound will not give the iodoform test if the CH_3CO- group is converted to acetic acid by hydrolysis. For example, acetoacetic acid gives a negative test. Ethanol, acetaldehyde, methyl ketones, and those secondary alcohols which yield methyl ketones on oxidation give a positive test.

$$CH_3CHO + 4NaOH + 3I_2 \longrightarrow CHI_3 + 3H_2O + NaOOCH + 3NaI$$

To perform the test, dissolve 100 mg of the unknown compound in 1 ml of water (dissolve in dioxane if insoluble in water). Add 3 ml of sodium hydroxide solution (10%) and then add dropwise a 10% solution of iodine in potassium iodide (20%) in water until a slight excess of iodine is present. Place the tube in a beaker of water at 60°C and add additional iodine until the brown color persists for at least 2 min. Then add dropwise aqueous sodium hydroxide (10%) until the brown color vanishes. Remove the tube from the water bath and add 10 ml of distilled water. Iodoform precipitates as a yellow solid, melting at 120°C. The compound can be satisfactorily chromatographed if the decomposition temperature of the iodoform is not exceeded in the instrument, as shown in Fig. 5-77.

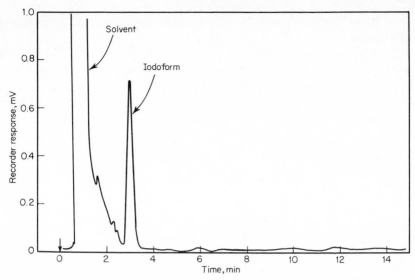

Fig. 5-77 Separation of iodoform on W-98 (10%) column (6 ft × ¼ in.) from 60 to 220°C.

Picric Acid

Addition Compounds. This reagent is useful in that it forms addition compounds with aromatic hydrocarbons, phenols, and phenolic ethers. Amines, almost all amino acids, hydrazines, thioureas, and alkaloids form picrates. Easily oxidizable compounds such as aliphatic aldehydes, reducing sugars, methyl ketones, hydrazines, and polyhydroxy phenols are oxidized by picric acid. Goswami et al[87] report that any compound containing a $>CH-$, $>CH_2$, or $-CH_3$ group contiguous to such negative groups as $-CHO$, $=CO$, $-NO_2$, or $-CN$ produces amber-red colors, indicating that reduction of the alkaline picrate solutions has taken place.

Molecular Compounds. *Aromatic hydrocarbons* such as benzene, toluene, ethyl and propyl benzenes, and xylenes produce unstable picrates that are difficult to prepare. The picrates of most of the other aromatic compounds can be prepared by the solvent or fusion method.[105] Prepare a saturated solution (2 to 3 ml) of the hydrocarbons in ethanol and a saturated solution of picric acid in ethanol. Mix the solutions in proportions so that equal molecular weights of the hydrocarbons and picric acid are present. Heat the solution mixture to boiling and cool. The resulting crystals are yellow, orange, or red in color.

The fusion method consists of mixing equal molecular amounts of the dry hydrocarbons and dry picric acid, heating to the melting point, and recrystallizing from ethanol. The melting point and the relative proportion of picric acid to the hydrocarbon are determined by chromatography (see Fig. 5-78).

To prepare *phenolic ethers* dissolve 1 μmol of the phenolic ether in the minimum required amount of boiling chloroform.   Similarly prepare 1 μmol of picric acid in boiling chloroform.   Mix the solutions and allow to crystallize.   Baril and Megrdichian[7] have reported the melting points and colors for 35 picrate derivatives of phenolic ethers.

To prepare *picrate salts* mix hot saturated solutions of the basic compound and picric acid so that there is a slight excess of picric acid.[134] Allow to cool and crystallize.   Water, 50/50 water-ethanol, pure ethanol, aqueous acetic acid (10%), and benzene can be used as recrystallizing solvents.

Picric acid as an oxidant can be utilized by placing 2 ml of saturated picric acid solution in a test tube and immersing it in a boiling water bath.   Add 1 ml of aqueous sodium hydroxide (5%) and 0.05 g of the unknown compound.   Keep in the water bath for 30 min.   If a reduction of the picric acid occurs, the yellow color will change to the red color of the picramate, which reduces one nitro group to an amino group.

Gas Chromatography of Picrates. Since many of picrates decompose in the instrument, the picric acid is converted to a trimethylsilyl derivative; then the ratio of the picric acid to the hydrocarbon can be measured and some indication of the molecular weight obtained, as shown in Fig. 5-79.

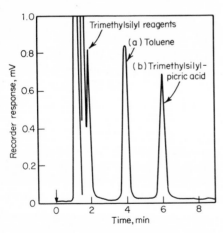

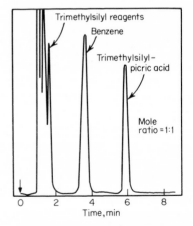

Fig. 5-78 Gas-chromatographic examination of hydrocarbon picrate as trimethylsilyl derivative, showing separation of (*a*) hydrocarbon and (*b*) picric acid on W-98 (10%) column (6 ft × ¼ in.) from 60 to 220°C at 10°/min.

Fig. 5-79 Ratio of picric acid to hydrocarbon as a measure of the complex (from Fig. 5-78).

Similarly, the trimethylsilyl derivative of the reduced picramate can be determined if the oxidized product is not volatile. Some indication of the oxidizing power of the sample can be observed by obtaining the ratio of picric acid to picramic acid and its relationship to the oxidized product, as shown in Fig. 5-80.

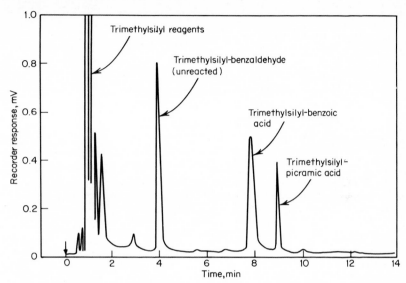

Fig. 5-80 Separation of picramate and oxidized product as trimethylsilyl derivative on W-98 (10%) column (6 ft × ¼ in.) at 60 to 220°C at 10°/min.

REFERENCES

1. Adams, R.: *Organic Reactions*, vol. II, chap. 8, John Wiley & Sons, Inc., New York, 1944.
2. Adams, R., and C. S. Marvel: *Org. Synth.*, collect. vol. 1, p. 504 (1941).
3. Arnold, R. T., and R. Lawson: *J. Org. Chem.*, **5**: 250 (1940).
4. Auger, V.: *Mikrochim. Acta*, **2**: 3 (1937).
5. Backeberg, O. G., and B. Staskun: *J. Chem. Soc. (Lond.)*, **1962**: 3961.
6. Baril, O. L., and E. S. Hauber: *J. Am. Chem. Soc.*, **53**: 1087 (1931).
7. Baril, O. L., and G. A. Megrdichian: *J. Am. Chem. Soc.*, **58**: 1415 (1936).
8. Battersby, A. R., and R. Binks: *J. Chem. Soc. (Lond.)*, **1955**: 2888.
9. Becher, P., and R. L. Birkmeier: *J. Am. Oil. Chem. Soc.*, **41**: 169–172 (1964).
10. Beckman Instruments: *Tech. Bull.* GC-90MI, figs. 5 and 6.
11. Bellaart, A. C.: *Rec. Trav. Chim.*, **83**: 718 (1964); *Tetrahedron*, **21**: 3285 (1965).
12. Benedict, S. R.: *J. Biol. Chem.*, **3**: 101 (1907); **5**: 485 (1908).
13. Biglow, H. R., and D. B. Robinson: *Org. Synth.*, collect. vol. 3, p. 103 (1955).
14. Billman, J. H., and J. W. McDowell: *J. Org. Chem.*, **27**: 2640 (1962).
15. Birch, A. J.: *Q. Rev.*, **4**: 69 (1950); A. J. Birch and H. Smith, *ibid.*, **12**: 17 (1958); G. W. Watt: *Chem. Rev.*, **46**: 317 (1950).
16. Blatt, A. H., and E. W. Tristam: *J. Am. Chem. Soc.*, **74**: 6272 (1952).

17. Bost, R. W., and F. Nicholson: *J. Am. Chem. Soc.*, **57**: 2368–2369 (1935); *Ind. Eng. Chem., Anal. Ed.*, **7**: 190–191 (1935).

18. Braude, E. A., and W. F. Forbes: *J. Chem. Soc. (Lond.)*, **1951**: 1762.

19. Broadbent, H. S., and C. W. Whittle: *J. Am. Chem. Soc.*, **81**: 1959: 3587.

20. Brown, H. C.: *Hydroboration*, p. 245, H. A. Benjamin, Inc., New York, 1962.

21. Brown, H. C., and P. Heim: *J. Am. Chem. Soc.*, **86**: 3566 (1964).

22. Brown, H. C., H. I. Schlesinger, and A. B. Burg: *J. Am. Chem. Soc.*, **61**: 673 (1939).

23. Brown, H. C., and B. C. Subba Rao: *J. Am. Chem. Soc.*, **82**: 681 (1960).

24. Buckler, S. A., L. Doll, K. F. Lind, and M. Epstein: *J. Org. Chem.*, **27**: 794 (1962).

25. Buckles, R. E., and C. J. Thelen: *Anal. Chem.*, **22**: 676 (1950).

26. Burgstahler, A. W.: *J. Am. Chem. Soc.*, **73**: 3021–3023 (1951).

27. Burchfield, H. P., and E. E. Storrs: *Biochemical Applications of Gas Chromatography*, pp. 299–309, Academic Press Inc., New York, 1962.

28. Burness, D. M.: *Org. Synth.*, collect. vol. 4, p. 628 (1963); R. H. Wiley and N. R. Smith, *ibid.*, 731.

29. Camera, E., and D. Pravisani: *Anal. Chem.*, **39**: 1645 (1967).

30. Carter, P. H., et al.: *Org. Synth.*, **40**: 16 (1960).

31. Chamot, E. M., and C. W. Mason: *Handbook of Chemical Microscopy*, 3d ed., vol. 1, John Wiley & Sons, Inc., New York, 1958.

32. Cheronis, N. D., and J. B. Entrikin: *Semi-micro Qualitative Organic Analysis*, p. 137, Thomas Y. Crowell Company, New York, 1947.

33. *Ibid.*, 2d ed., p. 365, Interscience Publishers, Inc., New York, 1957.

34. Clemmensen, E.: *Ber.*, **46**: 1837 (1913); **47**: 51, 681 (1914).

35. Coates, V. J., H. J. Noebels, and I. S. Fagerson: *Gas Chromatography*, pp. 73 and 87, Academic Press Inc., New York, 1958.

36. Cope, A. C., and E. Ciganak: *Org. Synth.*, collect. vol. 4, pp. 399 and 612 (1963).

37. Cope, A. C., et al.: *Org. Synth.*, collect. vol. 4, p. 218 (1963).

38. Cornforth, J. W., R. H. Cornforth, and K. K. Mathew: *J. Chem. Soc.*, 112–127 (1959).

39. Dakin, H. D.: *Org. Synth.*, collect. vol. 1, p. 149 (1941).

40. Dal Nogare, S., and R. S. Juvet, Jr.: *Gas-Liquid Chromatography*, p. 336, John Wiley & Sons, Inc., New York, 1962.

41. *Ibid.*, p. 407.

42. Davidson, D.: *J. Chem. Educ.*, **17**: 81 (1940).

43. Duke, F. R.: *Ind. Eng. Chem., Anal. Ed.*, **17**: 196 (1945).

44. Duke, F. R., and G. F. Smith: *Ind. Eng. Chem., Anal. Ed.*, **12**: 201 (1940).

45. Fatiadi, A. J., and W. F. Sager: *Org. Synth.*, **42**: 66, 90 (1962).

46. Feigl, F.: *Spot Tests*, p. 303, Nordeman Publishing Co., New York, 1939.

47. *Ibid.*, p. 272.

48. *Ibid.*, general reference.

49. *Ibid.*, p. 166.

50. Feigl, F., V. Auger, and O. Frehden: *Mikrochemie*, **15**: 183 (1934).

51. Feigl, F., V. Auger, and R. Zappert: *Mikrochemie*, **16**: 70 (1934).

52. Feigl, F., and O. Frehden: *Mikrochemie*, **16**: 79, 84 (1934).

53. Fieser, L. F.: *Organic Experiments*, p. 200, D. C. Heath and Company, Boston, 1964.

54. *Ibid.*, p. 177.

55. *Ibid.*, pp. 249–250.

56. *Ibid.*, pp. 719, 723.

57. Fieser, L. F.: *J. Biol. Chem.*, **133**: 391 (1940).

58. Fieser, L. F., and M. Fieser: *Organic Chemistry*, p. 215, D. C. Heath and Company, Boston, 1944.

59. Fieser, L. F., and M. Fieser: *Advanced Organic Chemistry*, p. 353, John Wiley & Sons, Inc., New York, 1966.

60. *Ibid.*, p. 779.

61. *Ibid.*, p. 890.

62. *Ibid.*, pp. 980 and 982.

63. *Ibid.*, p. 983.

64. *Ibid.*, p. 1081.

65. *Ibid.*, p. 204.

66. *Ibid.*, pp. 215 and 581.

67. Fieser, L. F., and M. Fieser: *Topics in Organic Chemistry*, p. 258, D. C. Heath and Company, New York, 1963; N. S. Hjelte: *Chem. Scand.*, **15**: 1200 (1961).

68. Fieser, L. F., and M. Feiser: *Reagents for Organic Syntheses*, p. 457, John Wiley & Sons, Inc., New York, 1967.

69. *Ibid.*, pp. 471–472.

70. *Ibid.*, p. 1289.

71. *Ibid.*, p. 723.

72. *Ibid.*, p. 890.

73. *Ibid.*, p. 35.

74. *Ibid.*, pp. 147 and 149.

75. *Ibid.*, p. 397.

76. Fieser, L. F., and M. Fieser: *J. Am. Chem. Soc.*, **56**: 1565 (1934).

77. Fischer, E., *Anleitung zur Darstellung organishies Präparat*, 8th ed., p. 39, Braunschweig, Germany, 1908.

78. Fischer, E.: *Ber.*, **22**: 2204 (1889); **23**: 930 (1890).

79. Freytag. H., and W. Neudert: *J. Prakt. Chem.*, **143**: 121, 180 (1932).

80. Fuson, R. C., and B. A. Bull: *Chem. Rev.*, **16**: 275 (1934).

81. Gaiffe, A., and R. Polland: *C. R.*, **252**: 1339 (1961); **254**: 496 (1962).

82. Gardner, J. H., and C. A. Taylor, Jr.: *Org. Synth.*, collect. vol. 2, p. 526 (1943).

83. Gillespie, H. B., and H. R. Snyder: *Org. Synth.*, collect. vol. 2, p. 489 (1943).

84. Gilman, H.: *Organic Chemistry*, vol. 3, chap. 1, and vol. 4, chap. 12, John Wiley & Sons, Inc., New York, 1953.

85. Gilman, H., and T. N. Goreau: *J. Am. Chem. Soc.*, **73**: 2939 (1951).

86. Gold, V., J. Hilton, and E. G. Jefferson: *J. Chem. Soc. (Lond.)*, **1954**: 2756.

87. Goswami, H. C., A. Shaha, and B. Mukergee: *J. Ind. Chem. Soc.*, **11**: 773 (1934).

88. Harrison, G. F.: in D. H. Desty (ed.), *Vapor Phase Gas Chromatography*, p. 336, Academic Press Inc., New York, 1957.

89. Hartman, W. W., and L. J. Roll: *Org. Synth.*, collect. vol. 2, p. 418 (1943).

90. Hearon, W. M., and R. R. Gustavson: *Ind. Eng. Chem., Anal. Ed.*, **9**: 352 (1937).

91. Herstein, B.: *J. Am. Chem. Soc.*, **32**: 779 (1910).

92. Hewlett-Packard Company: *Facts Methods*, **7** (3): 7 (June 1966).

93. Hochstein, F. A., and G. F. Wright: *J. Am. Chem. Soc.*, **71**: 2257 (1949).

94. Hodgson, H. Y., and E. W. Smith: *J. Chem. Soc. (Lond.)*, **1935**: 671.

95. Hoffman, J.: *Org. Synth.*, **40**: 76 (1960).

96. Holleman, A. F.: *Org. Synth.*, collect. vol. 1, p. 554 1(941).

97. Holleman, A. F.: *Rec. Trav. Chim.*, **23**: 169 (1940).

98. Holmes, R. R., and R. P. Bayer: *J. Am. Chem. Soc.*, **82**: 3454 (1964).

99. Hunt, T. H.: *Chem. Ind.*, **1961**: 1873.

100. Ingold, C. K.: *Structure and Mechanisms in Organic Chemistry*, p. 299, Cornell University Press, Ithaca, N.Y., 1953.

101. Jenner, E. L.: *Org. Synth.*, **40**: 90 (1960).
102. Jochims, J. C.: *Monatsh.*, **94**: 677 (1963).
103. Jones, R. G., and H. A. Shonle: *J. Am. Chem. Soc.*, **67**: 1034 (1945).
104. Jones, W. M.: *J. Am. Chem. Soc.*, **82**: 2528 (1960).
105. Jorgenson, M. J.: *Tetrahedron Lett.*, **1962**: 559.
106. Kahr, K., and C. Berther: *Ber.*, **93**: 132 (1960).
107. Kaiser, R.: *Gas Phase Chromatography*, p. 5, Butterworth Publications, Washington, D.C., 1963.
108. Kauffmann, H., and P. Panwitz: *Ber.*, **45**: 766 (1912).
109. Kirkland, J. J.: *Anal. Chem.*, **32**: 1388 (1960).
110. Kovache, A.: *Ann. Chem.*, (9)**10**: 184 (1918).
111. Kuhn, R., and W. van Klaveren: *Ber.*, **71**: 779 (1938).
112. Kuhn, W. E.: *Org. Synth.*, collect. vol. 2, p. 447 (1943).
113. Kuivila, H. G., and E. W. Walsh, Jr.: *J. Am. Chem. Soc.*, **88**: 571 (1966).
114. Kuivila, H. G., and O. F. Beumel, Jr.: *J. Am. Chem. Soc.*, **80**: 3798 (1958); **83**: 1246 (1961).
115. Kuivila, J.: *Org. Chem.*, **25**: 284 (1961).
116. Lange, N. A.: *Handbook of Chemistry*, 10th ed., p. 1277, McGraw-Hill Book Company, New York, 1961.
117. Leibnitz, H. C. E., and H. G. Struppe: *Handbuch der Gas Chromatographie*, Verlag Chemie Gmbh, Weinheim, Germany, 1967.
118. Leibrand, R. J.: *J. Gas Chromatogr.*, **5**: 518 (1967).
119. Leonard, N. J., and S. Gelfand: *J. Am. Chem. Soc.*, **77**: 3272 (1955).
120. Lucas, H. J.: *J. Am. Chem. Soc.*, **52**: 802 (1930).
121. McClure, J. D., and P. H. Williams: *J. Org. Chem.*, **27**: 24–26 (1962).
122. McKenna, J., et al.: *J. Chem. Soc. (Lond.)*, **1959**: 2502.
123. McReynolds, W. O.: *Gas Chromatographic Retention Data*, p. 145, Preston Technical Abstracts Company, Evanston, Ill., 1966.
124. *Ibid.*, p. 29.
125. *Ibid.*, pp. 36, 144, and 152.
126. *Ibid.*, general reference.
127. Martin, E. L.: *Org. Synth.*, **2**: 501 (1943).
128. Meerwein, H., G. Hine, H. Majert, and H. Sonke: *J. Prakt. Chem.*, **147**: 226 (1936).
129. Milas, N. A., and S. Sussman: *J. Am. Chem. Soc.*, **59**: 2345–2347 (1937).
130. Miller, A. E. G., J. W. Bliss, and L. A. Schwartzman: *J. Org. Chem.*, **24**: 627 (1959).
131. Minato, H., and T. Nagasaki: *Chem. Ind.*, **1965**: 899; *Chem. Commun.*, **1965**: 377.
132. Morgan, G. T., and F. M. G. Mickelwait: *J. Soc. Chem. Ind.*, **21**: 1375 (1902).
133. Morrow, C. A., and W. M. Sanstrom: *Biochemical Laboratory Methods*, 2d ed., pp. 157–166, John Wiley & Sons, Inc., New York, 1935.
134. Mulliken, S. P., and E. H. Huntress: *Manual of Identification of Organic Compounds*, p. 157, M.I.T. Bookstore, Cambridge, Mass., 1937.
135. Newbold, T. T.: *J. Org. Chem.*, **27**: 3919 (1962).
136. Newmann, M. S.: *J. Org. Chem.*, **26**: 582 (1961).
137. Newmann, W. P.: *Ann.*, **618**: 80 (1958).
138. Noller, C. R.: *Org. Synth.*, collect. vol. 2, p. 586 (1943).
139. Ochjai, E., and H. Mitarashi: *Chem. Pharm. Bull.*, **11**: 1084 (1963).
140. Overberger, C. G., J. G. Lombardino, and R. G. Hinskey: *J. Am. Chem. Soc.*, **80**: 3009 (1958).

141. Parsons, H.: *Symposium of Gas Chromatography.*, p. 88, Preston Technical Abstracts Company, Evanston, Ill., 1966.
142. Paul, H., and G. Zimmer: *J. Prakt. Chem.*, **18**: 219 (1962).
143. Pierce, A. E.: *Silylation of Organic Compounds*, p. 163, Pierce Chemical Company, Rockford, Ill., 1968.
144. Pietra, S.: *Ann. Chim. (Rome)*, **45**: 850 (1955).
145. Pino, P., and G. P. Lorenzi: *J. Org. Chem.*, **31**: 329 (1966).
146. Purnell, H.: *Gas Chromatography*, p. 371, John Wiley & Sons, Inc., New York, 1962.
147. Ramsey, J. B., and F. T. Aldridge: *J. Am. Chem. Soc.*, **77**: 2561 (1955).
148. Reeve, W., and J. Christian: *J. Am. Chem. Soc.*, **78**: 860 (1956).
149. Remick, A. E.: *Electronic Interpretation of Organic Chemistry*, John Wiley & Sons, Inc., New York, 1943.
150. Reppe, W., and H. Vetter: *Ann.*, **582**: 133 (1953).
151. Reusch, W., and R. Lemahieu: *J. Am. Chem. Soc.*, **86**: 3068 (1964).
152. Ritter, J. J.: *J. Chem. Educ.*, **30**: 395 (1953).
153. Roberts, G., and H. D. Green: *J. Am. Chem. Soc.*, **68**: 214 (1946).
154. Rothman, L. A., and E. I. Becker: *J. Org. Chem.*, **25**: 2203 (1960).
155. Rylander, P. N., and J. Kaplan: *Engelhard Tech. Bull.*, **2**: 48 (1961).
156. Schaefer, J. P.: *J. Org. Chem.*, **25**: 2027 (1960).
157. Schmidlin, J., and A. Wettstein: *Helv. Chim. Acta*, **46**: 2799 (1963).
158. Schultz, H. S., H. B. Freyermuth, and S. R. Buc: *J. Org. Chem.*, **28**: 1140 (1963).
159. Schwarz, R., and H. Hering: *Org. Synth.*, collect. vol. 4, p. 203 (1963).
160. Sherman, W. R., and E. C. Tayler, Jr., *Org. Synth.*, collect. vol. 4, p. 247 (1963).
161. Shriner, R. L., R. C. Fuson, and D. Y. Curtin: *Systematic Identification of Organic Compounds*, p. 123, John Wiley & Sons, Inc., New York, 1959.
162. *Ibid.*, pp. 136–147.
163. *Ibid.*, p. 160.
164. Smith, C. W., and R. T. Holm: *J. Org. Chem.*, **22**: 746 (1957).
165. Snyder, H. R., J. S. Buck, and W. S. Ide: *Org. Synth.*, collect. vol. 2, p. 333 (1943).
166. Soloway, S., and A. Lipschitz: *Anal. Chem.*, **24**: 898 (1952).
167. Stocker, J. H.: *J. Org. Chem.*, **27**: 2288 (1962).
168. Stoll, A., A. Lindemann, and E. Jucker: *Helv. Chim. Acta*, **36**: 268 (1953).
169. Swain, C. G., and C. B. Scott: *J. Am. Chem. Soc.*, **75**: 141 (1953).
170. Swamer, F. W., and C. R. Hauser: *J. Am. Chem. Soc.*, **68**: 2647 (1946).
171. Swern, D.: *J. Am. Chem. Soc.*, **69**: 1692 (1947).
172. Taborsky, R. G.: *J .Org. Chem.*, **26**: 596 (1961).
173. Taylor, E. C., Jr., and A. J. Crovetti: *Org. Synth.*, collect. vol. 4, p. 655 (1963).
174. Thiele, J., and F. Henle: *Ann.*, **347**: 307 (1906).
175. Todd, D.: *Org. React.*, **4**: 378 (1948).
176. Touda, K., et al.: *J. Org. Chem.*, **28**: 783 (1963).
177. Trahanovsky, W. S.: *J. Org. Chem.*, **31**: 2033 (1966).
178. Trahanovsky, W. S., and L. B. Young: *J. Chem. Soc. (Lond.)*, **1965**: 5777; *J. Org. Chem.*, **31**: 2033 (1966).
179. Treibs, W., et al.: *Ber.*, **86**: 616 (1953).
180. Trofimenko, S.: *J. Org. Chem.*, **29**: 3046 (1964).
181. Undenfriend, S., C. T. Clark, J. Axilrod, and B. B. Brodie: *J. Biol. Chem.*, **208**: 731 (1954).
182. Van der Kerk, G. J., and J. G. Noltes: *Appl. Chem.*, **9**: 106 (1939); **7**: 366 (1937).
183. Walther, R.: *J. Prakt. Chem.*, **52**: 141 (1895); **53**: 433 (1896).

184. Wasicky, R., and O. Frehden: *Mikochim. Acta.*, **1**: 55 (1937).
185. Waters Associates.: *Poropak Bulletin*, Framingham, Mass., 1966.
186. Watt, G. W.: *Chem. Rev.*, **46**: 317 (1950).
187. Weinberger, W.: *Ind. Eng. Chem., Anal. Ed.*, **3**: 365 (1931).
188. Welsbach Corp., Ozone Processing Division: *Organic Ozone Reactions and Technology*, 4th ed., January 1962.
189. West, E. F., and W. R. Brode: *J. Am. Chem. Soc.*, **56**: 1037 (1934).
190. Wheland, A. L.: *Organic Chemistry*, p. 502, John Wiley & Sons, Inc., New York, 1949.
191. Whitmore, F. C., and F. H. Hamilton: *Org. Synth.*, collect. vol. 1, p. 492 (1941).
192. Wilds, A. L.: *Org. React.*, **2**: 178 (1944).
193. Wilke, G., and H. Muller: *Ber.*, **89**: 444 (1956); *Ann.*, **629**: 224 (1960).
194. Williams, R. J., and M. A. Woods: *J. Am. Chem. Soc.*, **59**: 1408 (1937).
195. Wilson, W. J., and F. G. Soper: *J. Chem. Soc. (Lond.)*, **1949**: 3376.
196. Windaus, A.: *Ber.*, **39**: 2249 (1906).
197. Wiselogle, F. Y., and H. Sonneborn, III: *Org. Synth.*, collect. vol. 1, p. 90 (1941).
198. Wislicenus, H., and L. Kaufmann: *Ber.*, **28**: 1323 (1895).
199. Woodward, R. B., N. L. Wandler, and F. J. Brutschy: *J. Am. Chem. Soc.*, **67**: 1425 (1945).

Index

Index